BestMasters

Springer awards "BestMasters" to the best application-oriented master's theses, which were completed at renowned chairs of economic sciences in Germany, Austria, and Switzerland in 2013.

The works received highest marks and were recommended for publication by supervisors. As a rule, they show a high degree of application orientation and deal with current issues from different fields of economics.

The series addresses practitioners as well as scientists and offers guidance for early stage researchers.

Thomas Bentivegna

Innovation Network Functionality

The Identification and Categorization of Multiple Innovation Networks

Foreword by Prof. Dr. Frank Bau

 Springer Gabler

Thomas Bentivegna
Chur, Switzerland

ISBN 978-3-658-04578-4   ISBN 978-3-658-04579-1 (eBook)
DOI 10.1007/978-3-658-04579-1

The Deutsche Nationalbibliothek lists this publication in the Deutsche Nationalbibliografie; detailed bibliographic data are available in the Internet at http://dnb.d-nb.de.

Library of Congress Control Number: 2013955624

Springer Gabler
© Springer Fachmedien Wiesbaden 2014

Printed on acid-free paper

Springer Gabler is a brand of Springer DE.
Springer DE is part of Springer Science+Business Media.
www.springer-gabler.de

Foreword

Networks have always been used in any context. Politicians and careerists build their personal networks of power, students cultivate their Facebook network, and entrepreneurs build networks to find investors and customers as well as employees. In the last 20 years, formally structured and administered networks, cluster organizations, and the like have been established for many contexts. Regional developers and network administrators are proud of having the largest number of registered network participants and clicks on their internet platform. However, what ultimately counts are the real business contacts that lead to additional sales, sustainable supplier-relationships, or to innovation projects leading to sustainable competitive advantages for companies and regions. This still occurs through ad-hoc networks, i.e. networks that are informal and arise ad-hoc.

Thomas Bentivegna focuses in his Master Thesis on these ad-hoc networks, which are poorly represented in existing network and innovation literature. His work capitalizes on a collection of in-depth interviews conducted by a project team of the KARIM project, which is an INTERREG IV B project for the region of North West Europe run by the European Community. KARIM stands for Knowledge Acceleration Responsible Innovation Metanetwork. One of the actions taken in the project is to visualize ad-hoc innovation networks in an interactive tool allowing regional developers and SME managers to identify new approaches in running innovation projects and to learn from other companies and innovation support agencies in different regions. In his work, Mr. Bentivegna develops a methodology to cluster the qualitative interview data to finally identify seven types of ad-hoc innovation networks.

Prof. Dr. Frank Bau

Acknowlegements

I would like to begin this Master Thesis by thanking the people who aided in the design, preparation, and execution of my research. Without their help, none of this would have been possible.

At the very top of this list is my advisor and three-time professor, Dr. Frank Bau. His motivation, support, and constructive feedback is the foundation upon which this work is built. Any success which this Thesis generates can be traced back to the efforts of Dr. Bau.

Additionally, several other professor at the HTW Chur contributed valuable information which helped guide the research. Professor Dr. Ivan Nikitin gave helpful input concerning the methodology, while Professor Dr. Franz Kronthaler provided crucial support in designing some of the analysis methods contained herein.

A special thanks goes to Michael Forster for being my point-man in all things concerning KARIM and the data collected.

The firms which generously gave their time to the interviews and the KARIM interviewers must also be thanked, for without their effort there would have been no data to spend hours breaking down.

Finally, I would like to thank my wife for allowing me the freedom and space to work on this very time consuming, yet satisfying Master Thesis. By being by my side the whole time, her efforts played as large a role as any other in supporting this work.

Thomas Bentivegna

Table of Contents

List of Figures

List of Tables

Abbreviations

Abbreviation	Meaning
AG	Aktiengesellschaft (Publicly limited/held corporation)
CEO	Chief Executive Officer
CH	Switzerland
D	Germany
ENG	England
et al.	et alii (and others)
EU	European Union
F	France
GmbH	Gesellschaft mit beschränkter Haftung (Limited liability company)
HTW	Hochschule für Technik und Wirtschaft
KARIM	Knowledge Acceleration and Responsible Innovation Meta Network
IRL	Ireland
IT	Information Technology
KTT	Knowledge and Technology Transfer
MNC	Multi-National Corporation
NGO	Non-Governmental Organisation
OECD	Organisation for Economic Cooperation and Development
R&D	Research and Development
SIFE	Swiss Institute for Entrepreneurship
SME	Small and Medium Sized Enterprise
SPSS	Statistical Product and Service Solutions

1   Introduction

The first chapter of this Master Thesis outlines the background and starting position of this research project, as well as a description of the principle, KARIM. Moreover, the research problem, objectives, purpose, research questions, scope, and structure are presented.

1.1   Starting Position

The present climate of business places importance on applying a new management paradigm which has its central focus in a knowledge-based view, as opposed to an exclusive resource-based view. The older resource-based view holds that the competitive advantage which a firm seeks has roots in its capabilities to procure and adapt its human, physical, and organizational competencies in relation to its environmental surroundings (Barney, Wright, & Ketchen, 2001). Valuable and necessary resources for the creation of customer value must be in some form secured. On the other hand, newer knowledge-based approaches to management tend to view the problem of business in a different light. Here, the organization is considered to be a social community which excels in the creation and transfer of knowledge (Zander & Kogut, 1996). Ultimately, it is the management of this knowledge which leads to a competitive advantage for a firm. Due to the fact that the importance of knowledge in strategic management is becoming greater, new challenges are being created for today's managers attempting to cope with the intangible and elusive nature of knowledge transfer.

Common knowledge in business research dictates that firms must produce innovations in order to achieve and sustain competitive advantages, and the application of knowledge assets has become synonymous with effective innovation (Davenport & Harris, 2007). New innovations can be correlated with the creation of a competitive advantage in several ways, yet metrics for measuring an innovation's effect on a firm's performance can be a difficult task. One reality has become clearer in the last few decades - to create innovations efficiently and effectively, firms are more frequently operating in network structures (Cowan et al., 2005; Gloor, 2006, Pyka, 1999). The use of such networks by firms to increase their innovativeness has been steadily rising due to several reasons. Products and services are becoming increasingly complex due to the need to satisfy growing and varied customer demands. Essentially, this translates into an organization which must be able to integrate a wide set of competencies and skills. These complementary strengths are best found amongst partners in a network, as individual firms do best when they focus on their core competencies. Therefore, innovation networks become a logical option for advancing innovation. Additionally, firms often hedge their innovation activity risks while at the same time are given access to new knowledge, which often leads to new market possibilities (Penrose, 2008).

It can be assumed that network-based innovation is the cornerstone in the new logic of competition which is comprised of the diffusion of often complex knowledge and technologies as well as the gradually increasing pace of global technological change (Rycroft, 2007).

While the impact of technological advancement on business growth, progress, and development has long been accepted, the role and potential benefit of innovation networks has been widely ignored until recently. Schupeter, the famous early 20th century economist, first assigned the success of innovation to the individual concept of entrepreneurship in 1912 (Pyka, 1999). By the end of the second World War, economists identified specialized R&D laboratories of large firms and their routine-based, push-inspired new technologies, as the dominant form of innovation (Rothwell, 1994). It wouldn't be until the 1980's that the focus began to shift from the singular paradigm of innovation to the investigation of interrelatedness and interaction between all innovative actors - firms, universities, public research centers, and so forth. At present, the value-added created by these networks is a commonly held truth, and as such multiple public projects and government agencies globally have dedicated themselves to promoting and nurturing the use of such networks.

1.2 KARIM

The principle for whom this Master Thesis is designed to support is KARIM (Knowledge Acceleration Responsible Innovation Meta Network), a European Union INTERREG IV B project, which aims to facilitate knowledge transfer across North-West Europe with the ultimate goal of increasing overall economic competiveness in the region. The project KARIM, a multinational effort, is supported by eight partners from seven different countries. Financing for the project is done through the European Regional Development Fund (INTERREG IVB North-West Europe). The aims of the project are (Bau, 2011):

- to improve small and medium sized enterprise's, or SMEs, access to innovation encouraging/promoting organizations.
- to establish a network of at least 500 innovation encouraging/promoting organizations.
- to internationalize the encouragement/promotion of knowledge and technology transfer between universities/public institutions and SMEs.
- to increase the cooperation and connections between international and regional systems of innovation.
- to establish a level of influence over innovation policy through concrete results of this Master Thesis and other projects.

KARIM is represented in Switzerland by the University of Applied Sciences HTW Chur and led by Prof. Dr. Frank Bau, the head advisor to this Master Thesis.

1.3 Research Problem

KARIM strives to identify existing innovation networks, aid and encourage cooperation and the use of such networks, and create policy recommendations to support these networks. (Bau, 2011). Common knowledge and theory leads to the assumption that these networks exist and are aiding firms from the region to supplement their innovation activities. The problem lies in the fact that currently these innovation networks have as yet not been identified and catalogued in detail within the project borders of KARIM.

1.4 Research Objectives

Based on the research problem, the main aim of this research is the presentation of a full report analyzing existing innovation networks from the region of interest to KARIM. This entails a full illustration of the innovation networks themselves and an understanding of the dynamics through which they function. Specifically, two main objectives were created at the onset of this research:

1. The creation of a detailed evaluation of innovation network activities operating within the jurisdiction of the KARIM project region, including a description and characteristic analysis of several distinct types of networks - this essentially provides a **network perspective**.

2. A closer examination of the firms operating within these identified networks and what similarities bind these firms together within the same network structure - this essentially provides a **firm perspective**.

1.5 Research Purpose

The purpose of this Master Thesis is to help support KARIM project leaders in achieving their goals in making decisions which could have policy ramifications in terms of supporting these, and other, innovation networks, solidifying the practical application of this report as an information reporting vehicle. The information presented in this Master Thesis can secondarily be used to increase the base of knowledge concerning the practical use of innovation networks, especially within the region of North-West Europe. Additionally, other firms may benefit from learning about where existing innovation networks are located by motivating them to operate using a more open style of innovation. A full list of the practical implications of this research can be seen in Chapter 7.

1.6 Research Questions

Consistent with any research, questions provide a good foundation to build upon to create and structure the direction of a report. In order to accomplish the objectives of this Master Thesis, two central and paramount questions were identified and answered:

1	*What types of innovation networks exist and are being used within the KARIM project region?*

2	*What are some general characteristics of firms using these individual innovation networks?*

The questions were initially left purposely general in order to keep a level of flexibility within the project and discourage any conscious tampering of the data, which is sometimes done to answer a more specific question or set of questions. These questions are a direct interpretation of the Research Objectives, as question 1 represents the first objective, or network perspective, while question 2 represents the second objective, or firm perspective.

1.7 Research Scope

The scope of research contained within this Master Thesis is limited to firms operating in the geographical region of North-West Europe, which includes Germany, Switzerland, France, England, and Ireland. The sample framework for firm selection is discussed in detail in Chapter 3.3.1. Moreover, the delimitations of the research scope are further explained in Chapter 4.5.

1.8 Research Structure

This Master Thesis is linearly structured according to accepted and logical research practices. Chapter 1 is the introduction, which provides a starting point and background information for the problem, as well as critical reasons why the research is being done and the major objectives contained herein. Chapter 2 is a presentation of the relevant literature/theory through other studies concerning innovation networks and knowledge transfer in order to provide the status-quo of currently accepted theory, along with identifying a research gap. Chapter 3 outlines the methodology and explains how the empirical research was conducted by describing the entire research framework. Chapter 4 covers various aspects of the research process, including validity and reliability issues of this report, as well as limitations and other ethical considerations. Chapter 5 and 6 present the results of the empirical research, while Chapter 7 provides a listing of the theoretical and practical implications, as well as future possible research goals for the KARIM project based on these results. Figure 1 on the following page shows an illustration of this structure.

Figure 1: Research Structure

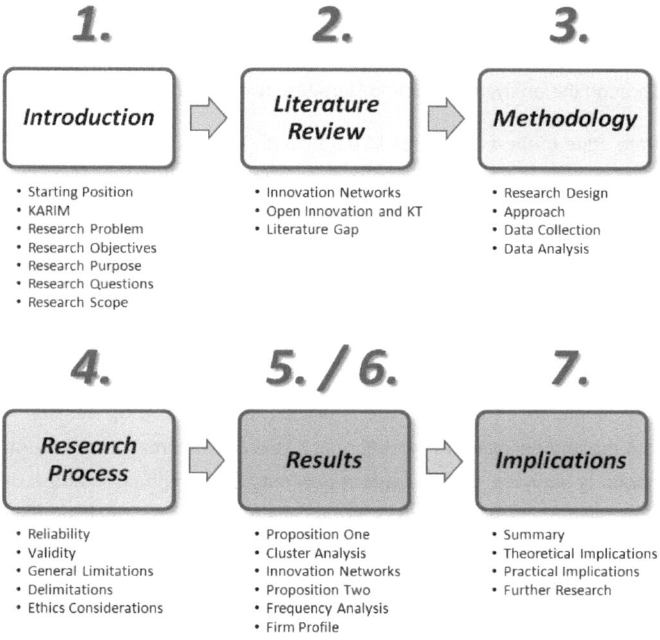

Source : Own Illustration

In addition to the above mentioned structure, a full appendix accompanies this research. The appendix consists of supporting material to that which is discussed within the report. This material includes a list of all innovation actors mentioned by the interviewed firms, Data Matrixes for both the network and the firm perspective, a full graphical listing of all specific innovation networks found within KARIM's Data Set, and the interview guide used by KARIM interviewers in preparing and conducting their firm interviews. Moreover, a full list of the quoted sources is given in the References.

Finally, a CD of additional data is also included with this report. This CD includes the raw data which was used to create this Master Thesis (KARIM's Data Set and all available transcribed interviews), the cluster analysis variables, and this report in a digital format along with a one page summary.

2 Literature Review

As a central feature of this Master Thesis, a literature review of two main topics was conducted. The two main topics researched were Innovation Networks and Knowledge Transfer. Through the analysis of existing literature, two objectives were pursued:

1. To illustrate state-of-the-art theories in both fields. This includes an extensive analysis of the definitions, motivations, benefits, and barriers to firms in both areas of research.

2. To identify gaps in the existing research and where this study can fill those gaps and contribute to the general discussion in both areas of research.

2.1 Innovation Networks

Definition

The definition of innovation is one which is difficult to come to a consensus. However, the OECD's Oslo manual defines an innovation as "...the implementation of new or significantly improved goods or services, or a process, a new marketing method, or a new organizational method in business practices, workplace organization, or external relations (OECD-2, 2005, pg 46)". An important distinction of innovation is that it must create some value for the organization which implements it.

Studies conducted in the 1960's and 70's first highlighted the relative importance of external sources of inputs into an organization's innovation processes (Gibbons & Johnston, 1974). As a result of this trend, innovation networks have increasingly replaced roles and functions traditionally undertaken within the closed confines of a firm. An examination of existing theories provides three varying yet similar definitions to the concept of innovation networks:

1. Innovation networks are defined as a set of actors who are interconnected by a series of relationships which ultimately is targeted towards the creation of a new innovation (Busquets, 2010).

2. A innovation network is a social network which consists of a finite set or sets of actors and the relation or relations defining them (Wasserman & Faust, 1994).

3. Networks of organizations and people can be described as innovation networks when competitive advantages are realized through the activities of these networks. The main functions found within any innovation network revolve around innovation, collaboration, and communication (Gloor, 2006).

4. Innovation networks are "...real and virtual infrastructures and infratechnologies that serve to nurture creativity, trigger invention, and catalyze innovation in a public and/or private domain context, for instance, government-university-industry, public-private research and

technology development, co-opetive, which is a combination of cooperation and, sometimes, competitive partnerships (Carayannis & Campbell, 2006, pg. 8)."

Role of Globalization

The proliferation and expanded use of such networks might be linked to multiple factors, however a paramount factor at play is perhaps increasing globalization. Globalization is typically identified with pressures that have resulted in cooperation among various types of innovative organizations. The inherit process of globalization increases the value of access to unique knowledge which is located in different areas and countries, leading to a need to form deep linkages and cooperation in such innovation networks. Perhaps less subtle is the role which innovation networks play themselves in advancing globalization, as cooperative innovation produces complex and overarching relationships which help form global markets. It can be theorized that the advancing regionalization/globalization of consumer markets will only serve to underscore the importance of such networks, while the pointed use of these networks will become a driver of global interconnectedness (Rycroft, 2007).

In first-world developed countries, the long-term growth of firms and, therefore, the region, sprouts from an innate ability to continually develop innovative products. The competencies needed to innovate require access to "invisible factors", or tacit knowledge coupled with sticky information (Hippel, 1994), which can be difficult to procure, but are more easily accessed through linkages in networks. This underpins the theory that globalization increases regionalization, since knowledge and know-how are often found on a global stage and generally utilized in a local and/or regional context. Additionally, the increasing specialization of organizations and firms which allows them to seamlessly integrate into an innovation network are the same qualities seen in driving global business (Sternberg, 2000).

Success Factors

A fully functioning innovation network can be comprised of a multitude of actors - (internally and externally) and a diversity of relationships (formal and informal). Regardless of its composition, size, or scope, Hopkins identifies 5 major criteria, or success factors, which must be satisfied for the network to function properly (Hopkins, 2003):

- Knowledge creation, utilization, and transfer: a major goal of innovation networks is to create and disseminate knowledge.
- Dispersed leadership and empowerment: highly effective innovation networks are comprised of capable people who collaborate and work well together. The skill-set needed for the proper functioning of an innovation network are related to those needed for the operation of effective teams.

- Adequate resources: Time, finance, and human capital are central to a network meeting its goals.
- Clarity of structure: Effective networks are well maintained and organized with logical and clear operating procedures and mechanisms for ensuring that proper participation can be achieved.
- Consistency of values and focus: It is crucial that innovation networks have a common and clear aim/purpose. The values which underpin the network must be well communicated and shared by all.

The degree to which an organization can benefit from any knowledge flow which occurs is heavily dependent on the absorptive capacity of the individual unit. The absorptive capacity refers to the ability of the organization to easily integrate external knowledge into its own working stock. The channels of knowledge flow represent one of the most important advantages to using an innovation network. Conversely, absorptive capacity generally increases with the organization's prior cooperation experiences, which highlights the importance of the repeated use of such networks (Küppers & Pyka, 2002).

Benefits of Innovation Networks

An important consideration associated with innovation networks is determining what value which they provide its members and how they generate worth for all involved. Existing theory has covered this extensively and the results are varied and diverse. In general, individual firms have five motives which drive them to participation in such networks: The high costs and risks of R&D projects, an attempt to shorten the period between discovery and market introduction, the exploration potential of new markets and new market niches, technology transfer, and the monitoring of technological opportunities and evolution. (Pyka, 1999). As new technologies have been introduced and integrated into current product spaces at rapid rates, the need to mesh new kinds of expertise and knowledge - both in production and innovation activities - has created a new demand previously unseen. This widening of the technological base creates a gap in firms in that the information and technology needed for innovation may lie outside their traditional core competencies, creating motives to enter and use innovation networks. This often leads to firm's filling structural holes, increasing their position of centrality, and the formation of regional temporary clusters. (Cowan, Jonard, & Zimmermann, 2005).

Taken as a whole, it can be said that that any firm which utilizes the services offered by an innovation network can gain a competitive advantage by capitalizing on various organizational benefits, as show on the next page in Figure 2. In general, innovation networks help a firm to achieve these competitive advantages by:

Figure 2: Organizational Benefits of innovation Networks

Organizations are more innovative and collaborative ready
Organizations are more ambidextrous
Hidden business opportunities uncovered before competition
Synergies can be released in firm
Organizations can locate experts
Organizations can become more secure

Gains to a Competitive Advantage

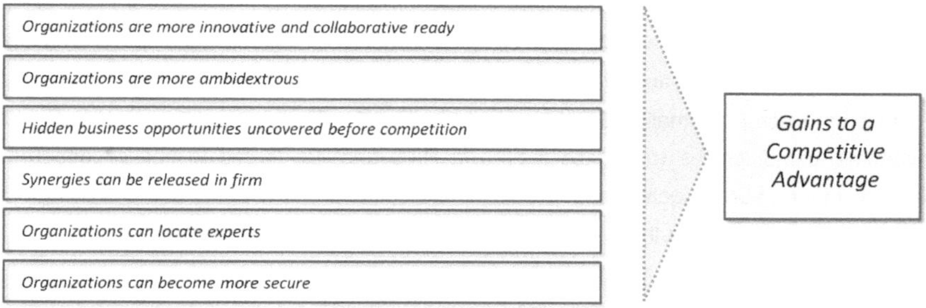

Source : Own Illustration based on (Gloor, 2006)

1. Making organizations more innovative and collaborative: by exposing itself to other capabilities not concentrated on within the confines of a firm, the intrinsic innovation capabilities of a company can be improved, leading to a better overall culture.

2. Making organizations more ambidextrous and able to react to market and technology changes as well as new and disruptive market entries: disruptive innovations can enter a marketplace and change the value proposition of a product's paradigm. Firms which are plugged into networks of collaborators often are able to identify threats quicker or themselves become the producer of a disruptive innovation.

3. Uncovering hidden business opportunities before the competition: new markets and distribution channels often come about as a result of innovation networks complimentary processes.

4. Allowing for the release of synergies within the firm: collaborating is a skill which often must be trained with employees. By actively participating in external innovation activities, firm employees will be more likely to cooperate internally as well.

5. Helping organizations locate experts: business problems require solutions which are not always available in-house. Theory holds that if a firm cannot find experts within the confines of a particular innovation network, it will be more likely to find an expert elsewhere based on the contacts it has made within the initial network.

6. Leading to more secure organizations: long-term prospects of a firm are vastly improved by engaging in such activities.

In addition, the various actors of an innovation network generally look to accomplish several goals together as a unit: sharing the costs and risks of innovation (Penrose, 2008), taking advantage of the transaction costs (Williamson, 1991), using strategic and resource advantages, benefiting from the trust in a network dynamic (Van de Ven, Polley, Garud, &

Venkatraman, 2008), and benefiting from the soft social and cultural fit among the actors (Kogut, 2000).

Another important aspect which must be considered when examining a firm's innovation network activities is the financial flow advantage which could be obtained as opposed to merely innovating internally. Start-ups and developing firms generally are missing the economic competencies to finance their own R&D and are forced to find a cooperation partner. In the case of such partnerships, the intellectual property rights of the ensuing R&D results must be defined at the beginning of any undertaking (Küppers & Pyka, 2002).

As an added feature of using networks to innovate, organizational learning can be essentially better addressed and articulated as a result. Firms embedded in an innovation network are required to not only know what other firms/organizations are doing, but how they are doing it as well. To support this inter-firm learning process of long-range know-how, a fixed and cooperative environment is necessary (Pyka, 1999). This points to an improvement not only in the firm as a whole, but in individual efforts as well. Individuals in a network are able to build a wider base of colleagues by being in contact with committed people with shared values. Moreover, personal relationships with leaders of a field can be established. Individuals who participate in their firm's activities within a network often learn new skills and become experts themselves (Gloor, 2006).

The range of the benefits gained through the use of innovation networks is indeed broad and transcends the boundaries of any single firm. Companies can generate momentum in growth both from and for their region, as innovative linkages hold a critical importance for many businesses scattered throughout multiple regions. In the absence of such linkages, lock-in effects can occur which lead to stagnation and path dependency. Simmie and Kirby identify innovation networks as "...the nexus between the forces which increasingly expose more firms to international, competitive markets and the structures and strategies which they adopt to compete successfully in those markets and the local conditions, regulations, and regimes which enable parts of or whole firms to export competitively into those international markets. (Simmie & Kirby, 1998, pg. 19)." This underscores how externalities which positively occur in a firm due to its activities related to an innovation network can indeed benefit the region in general.

Innovation Network Members

Examined from an academic perspective, it can be said that innovation networks are comprised of members who assume one of 5 roles (Van de Ven, Polley, Garud, & Venkatraman, 2008):

- The sponsor who procures, advocates, and helps advance innovations.

- The entrepreneur who manages innovations and ventures by often breaking the traditional mold of innovation activities.
- The critic who challenges and questions investments, goals, and progress.
- The mentor who supports processes of innovation by coaching and counseling.
- The institutional leader who often brings all parties together as an intermediary.

When approached from a more pragmatic position, the different actors of a viable innovation network form 6 major categories of players. These categories comprise the bulk of typical actors (Conway & Steward, 1998):

1. Upstream actors : suppliers, producers, etc.

2. Downstream actors : customers, handlers, distributors

3. Competitive actors from rival firms : firms both within and outside the same industry

4. Knowledge generating actors : universities, research institutes, etc.

5. Regulatory actors : government bodies and trade unions

6. Social environment actors : pressure groups, NGOs, etc.

A network can conceivably consist of any constellation of these different actors providing one or more of the above mentioned roles. A relatively important concept discussed within the existing literature of innovation networks focuses on the strength of ties between the various actors (Gilseng & Nooteboom, 2004). Central to understanding this is the recognition that such ties can be strong and cohesive, which aids in cultivating trust and collaboration between members. Conversely, these ties can also present themselves as weak and thin, which would allow for diversity, flexibility, and lost cost exploration. One side of the argument postulates that dense and strong ties aid in creating social control and build up reputation and social capital, which in turn helps foster a spirit of cooperation. The other side states that weak ties, rather than strong, are appropriate in order to gain access for new and simple information. Often, in networks where frequent and intense interaction between actors occurs, much of the information circulating within the system is redundant. It has been suggested that a combination of the two different types of intensity between actors would be ideal, as they serve to provide two different kinds of knowledge: strong ties help promote and transfer complex knowledge, while weaker ties are aimed towards transferring more simple and less coordinated knowledge (Uzzi, 1997).

The relationship between the geographical proximity of actors within a network and its ultimate effectiveness is an issue which has been explored and researched. Technological districts, technopoles, and science parks have all been seen as not only sparks igniting the creation of informal networks, but also as incubators where knowledge creation occurs at

12

faster rates than otherwise. More recent works in the field of economic geography outlines the importance of spatial consequences in technological spill-overs between academic and industry partners (Audretsch & Feldman, 1996).

While geographical proximity is an important concept in facilitating the transfer of knowledge between organizations, the concept of internal organizational proximity may be just as, if not more important to the underlying success of innovation networks. Organizational proximity is characterized by the collective value systems, or the company culture, which tends to homogenize individual behaviors within that organization. This collective culture aids in employees being able to spontaneously and accurately interpret exchanged knowledge the same way. Geographical proximity is effective only when it co-exists with internal organizational relationships. It appears that organizational proximity is a stronger supporter of technology transfer and innovation diffusion than geographical proximity alone (Rallet & Torre, 2000).

A generic view of how an innovation network can function is presented in Figure 3, showing the various actors and their respective roles towards fulfilling their network goals.

Figure 3: Innovation Network Actors

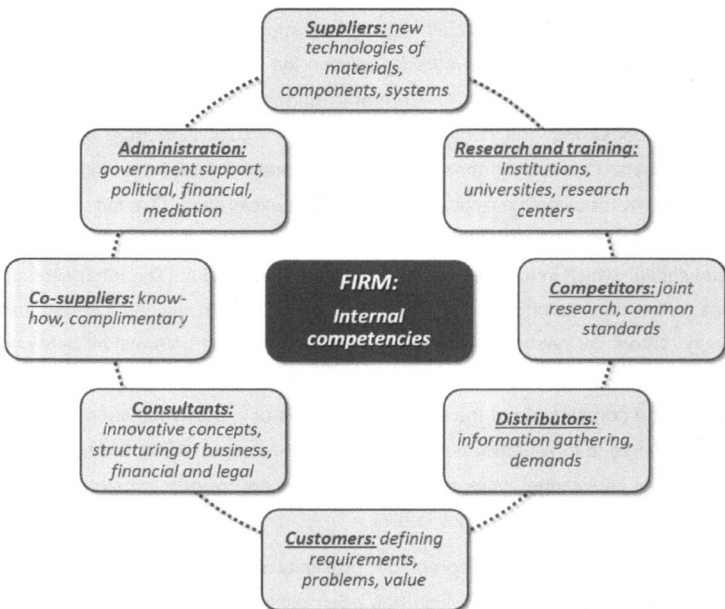

Source : Own Illustration based on (Pittaway, Roberston, Munir, Denyer, & Neely, 2004)

Frameworks of Innovation Networks

There are several different frameworks/structures within which innovation networks can function: for ecample, national vs. regional and formal vs. informal. As a starting point, national systems of innovation can be described as a system of structured interactions between agents who are involved in the process of generating technological progress on a national scale within a country (Pyka, 1999). In this framework, firms and public research institutions such as universities are typically key drivers to the network's activities. Regional structures can be very similar in terms of their activities and member roles, but the positive externalities to the region (improved employment, increased innovativeness of surrounding businesses, etc.) which result is the deciding factor which separates them from a merely national structure. Silicon Valley, Route 128, Wissenschaftsstadt Ulm, and Emilia Romagna are examples of innovative regions which foster regional networks of cooperation, resulting in an overall increase in efficiency for the area. Often, firms which have participated within the confines of a regional network are better prepared to branch out to more national/global institutions, based largely on the experiences and contacts which these firms were able to locally develop (Cooke & Morgan, 1994).

Formal innovation networks can take several forms: Joint R&D agreements, direct investment, licensing of technology, the buildup of a common R&D infrastructure, or research associations (Freeman, 1991). Generally following the course of contractual stipulations and often regulatory and statutory considerations, these types of networks are single-minded in goals but often rigid in terms of the flexibility afforded the members. Recent literature focuses, however, on the growing importance of informal networks. Formal contracts are increasingly becoming displaced by less rigid informal relationships. Behind most formal networks are various types of informal, or ad-hoc, frameworks. These are more fluid in how they operate and allow for a more dynamic relationship between actors which can, and often does, change over time. Although seldom systematically measured, informal networks appear to be important due to their multiple sources of information and diverse patterns of collaboration (Pyka, 1999).

The increase in alliances which are agreed upon to gather or exchange knowledge has helped to break down the traditional roles of markets and firm hierarchies. Networks are different from these hierarchies in a multitude of ways, but primarily in that they depend on particular types of interactions between the actors located within the network. Within a network, company borders are more porous in nature and firms thrive by having relationships with others who have complimentary assets. These relationships between actors can be defined as (Cowan, Jonard, & Zimmermann, 2005):

- Relational embeddedness: The relationship between actors is purely of a contacting/networking nature.
- Cognitive embeddedness: The relationship between actors is of a knowledge sharing/creation nature.
- Structural embeddedness: The relationship between actors takes advantages of organizational structures or processes which one partner has and the other requires.

The different resources of individual partners are combined in any network scheme, thus determining the complementary nature of their relationships. When two or more organizations innovate jointly, the process will be most successful when knowledge profiles complement one another. If two firms' knowledge profiles are too close together, there will be too great of an overlap resulting in little reason to share. If the knowledge profiles are too far apart, difficulties may arise in partners' understanding of one another (Grant, 1996) (Nooteboom, 1999).

Problems with Innovation Networks

As with any system of interaction, innovation networks are not immune to problems which could result in their effectiveness being undermined or even prevent their forming in the first place. There are 4 main areas of concern when dealing with multiple partners in a collaborative effort, as shown in Figure 4 (de Man, 2008):

Figure 4: Areas of Concern with Multiple Partners

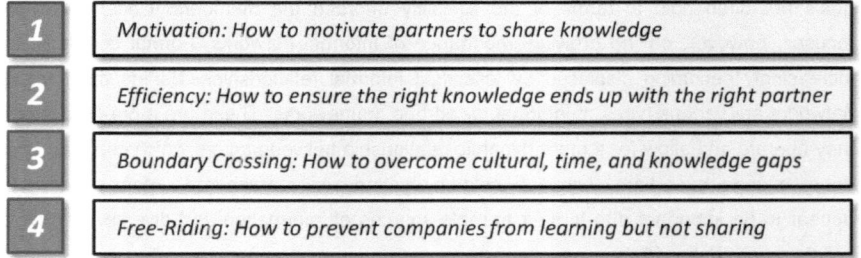

Source : Own Illustration based on (de Man, 2008)

1. Motivation: How to motivate partners to share knowledge: There must be incentive for all parties involved to transfer knowledge to others who require it. In the absence of incentive, there can be no transfer.
2. Efficiency: How to ensure the right knowledge ends up with the right partner quickly: Possessing and being motivated to transfer knowledge is the foundation upon which proper use of that knowledge is laid. The identification of gaps in competency among the members of an innovation system is key.

3. Boundary Crossing: How to overcome cultural, time, and knowledge gaps. Once gaps are identified in all areas, action plans must be developed which can help accomplish this. Formal networks with clear leadership roles have distinct advantages here over more ad-hoc, free-spirited networks.

4. Free-riding: How to prevent companies learning but not sharing: By illustrating the intrinsic benefits to all when everyone participates, knowledge hoarders can be coaxed into understanding how their proactive role in the network is just as important as their reactive role.

In general, issues can arise which result in the failure of the entire innovation system as a whole. This includes the thinness of an organization, or a lacking of the proper capabilities needed by other members of the network. Additionally, lock-in, or being such a specialized member of the network that any knowledge which can be offered would be effectively useless to the others, is also a factor to be considered. Fragmentation, or a breakdown in the dynamic interaction between actors for various reasons, can also lead to a system failure (Tödtling & Trippl, 2005).

More pessimistic economists and business researchers note that whatever the objectives of collaboration are, the result is often a failure, as strategic alliances have a success rate of less than 50% (Faems & Van Looy, 2003). Networks themselves are not immune to common conflicts, disputes, and lack of coordination which dooms many internal innovation projects. Similarly, there is some evidence that suggests faster commercialization of a product does not necessarily result in commercial success (Meyer & Utterbeck, 1995). Moreover, cooperation does not always correlate to enhanced innovation speed. Complex innovations may be hindered due to the fact that combining and synthesizing knowledge bases related to multiple systems of a product is difficult to quickly diffuse and communicate through a network (Rycroft, 2007). However, such research outlining the negative effects of networks is minimal compared to the overwhelmingly large amount of positive benefits obtained by firms in using such networks.

2.2 Knowledge Transfer

Definition

The ideals espoused by open innovation have been present in economic theory as far back as the 1960's, though often the term is associated with modern principles in competitive business practice. Central to any understanding of knowledge transfer is a review of the definition of open innovation. Henry Chesbrough, arguably the most prolific promoter of the concept, identifies the term open innovation as:

"Open innovation is a paradigm that assumes that firms can and should use external ideas as well as internal ideas, and internal and external paths to market, as the firms look to advance their own technology. It is innovating with partners by sharing risk and sharing reward (Chesbrough, 2003, pg. 4)."

The key to an open innovation analysis is the realization that the flow of knowledge is the deciding factor between an open or closed system. Expanding upon the basic framework of his original definition, Chesbrough et al. add that open innovation occurs by "...the use of purposive inflows and outflows of knowledge to accelerate internal innovation and expand the markets for external use, respectively. (Gassman, Enkel, & Chesbrough, 2010, pg. 1)."

Knowledge transfer, in being a crucial aspect of open innovation, is seen as a process through which a firm can recreate complex and casually ambiguous sets of routines in new forms and continuous functions, thereby instilling a sustainable aspect to the term (Szulanski, 1999). In fact, knowledge creation can been said to be a crucial source of sustainable competitive advantage, and this creation is often accomplished through using outside sources (Osterloh & Frey, 2000). Knowledge in general is said to be sticky in nature, that is, difficult to transfer and be assimilated into a new organization. Because of this, important considerations to be made when dealing with the transfer of knowledge include the various kinds of contextual variables along with the process through which transfer implementation occurs.

Knowledge transfer can be divided into two main parts: internal and external. The most basic definition of internal knowledge transfer is the dissemination of knowledge from one division to another division within the same firm (Lord & Ranft, 2000). In other words, the internal transfer of knowledge can be viewed as all the processes used to make knowledge available about all organizational routines to its members. External knowledge transfer takes on a similar character, however the knowledge transfer occurs across company lines. Here, the exchanging of information about managerial practices and performance outcomes often takes precedence, with a particular focus on an exchange of content specific knowledge (McEvily, Das, & McCabe, 2000). Therefore, boiled down to its bare essentials, external knowledge transfer is perceived as acquiring existing knowledge form external sources through the crossing of a firm's borders.

Often, knowledge is closely associated with competence development within a firm. The development of competence in an organization invariably leads to some type of internal change. Here, three levels of change are relevant: invention and modifications of invention activities, innovation and change processes, and organizational change capabilities. All of these levels of change within a firm are supported by the individual skills possessed by members of the organization, structures and processes, incentives, and corporate culture

(Carayannis & Campbell, 2006). The changes which occur to these levels may be incremental or radical in nature, however, of the upmost importance is the recognition that these changes only occur in the presence of new knowledge.

Motivation of Firms

Firms can be motivated in different ways to attempt external knowledge transfer processes with other organizations, and thereby participating within an open innovation framework, for several reasons, as shown in Figure 5 (Arvantitis, 2009):

Figure 5: Motivation to Innovate Openly

Source : Own Illustration based on (Arvantitis, 2009)

The amount of positive external knowledge which is transferred into an organization, otherwise known as incoming spillovers, is closely related to the absorptive capacity within a company. Conversely, outgoing spillovers refer to the amount of a firm's knowledge which flows out of a firm and is able to be used by others. Incoming spillovers motivate a company to attempt an R&D cooperation, while outgoing spillovers effect an opposite reaction by stopping innovation activities due to the risk of sensitive internal knowledge falling into competitor's hands (Cassiman & Veugelers, 2002). Strategically speaking, firms attempt to limit their exposure of outgoing spillovers through secrecy measures or by being involved in overly complex projects. However, all firms essentially must make the determination that incoming spillovers' positive effect on the firm is usually greater than the negative aspects stemming from outgoing spillovers.

There are several factors which lead to a firm's decision regarding with whom to cooperate. A firm is more likely to cooperate and engage in knowledge transfer activities with a university when appropriability and quality competition are important driving forces. With university cooperation, firms try to realize synergies in innovation activities and they take care about outgoing spillovers which might increase product market competition. On the other hand, a firm is more likely to cooperate with a competitor, known as horizontal cooperation, when the value of collusion matters, as there is an inverted U-shape relationship between the number of principal competitors and the probability of cooperation among competitors. This indicates that in markets with a medium number of principle competitors, horizontal cooperation is more likely compared to markets with very few competitors and markets with many competitors. Competition authorities in government are likely to prohibit cooperation in markets with very few competitors and in markets with many competitors the gains from horizontal collusion are diminishing (Bolli & Wörter, 2011).

Another aspect to motivation are firm's incentives to interact with public science institutions in order to begin a process of knowledge and technology transfer (KTT) with the purpose of gaining new tacit and/or codified scientific knowledge in research fields which are relevant for their own innovation activities. A study of Swiss firms indicates that firm size and industry play a role in determining a predisposition towards cooperation. Of the researched firms, 25% of small firms were involved in KTT activities compared to 47% of large firms. Firms in high-tech manufacturing and in knowledge-based services showed the highest tendency to be involved in KTT, especially concerning firms in the chemical, electronics, and business services industries (Arvanitis, Kübli, & Wörter, 2005).

Since science institutions and academic partners are often involved in knowledge transfer activities, existing theory is abundant in terms of identifying what concrete motives a firm usually has in deciding to become involved with them (Arvanitis, Kübli, & Wörter, 2005):

1. Access to human capital: This can be subdivided into the following areas:
    a. Access to abilities in addition to internal know-how
    b. Further education and training possibilities
    c. Project characteristics requiring cooperation
    d. Insufficient firm R&D resources
2. Financial motives: Access to funding to alleviate project costs
3. Access to research results: Binding of theory and praxis experience

Benefits of Open Innovation and Knowledge Transfer Activities

Intrinsically, firms use tactics associated with open innovation to close various gaps which may exist within the company. When the determination is made that these gaps cannot be overcome through traditional internal processes, incoming spillovers have proven to be beneficial (Howard-Partners, 2007). There are several types of gapes which a firm may potentially have. Information gaps are gaps encountered by firms in identifying relevant, useful, and applicable techniques for product and service development. Access gaps refer to difficulties which a firm has in accessing technologies and knowledge which they know exist but are unsure about how to acquire it. Transfer gaps are gaps which surround the negotiation of license and consultancy/contract agreements by using external intermediaries, as well as project management which may be beyond the competency level of the business, especially in the case of SMEs. Translation gaps points to developing and transforming knowledge embedded in a technology into a form and format which can be used in a product, service, and/or business development.

In general, firms which cooperate see a positive impact on their knowledge absorptive capacity, meaning they are better able to identify ideas and processes from outside sources which benefit the company the most. Additionally, cooperative R&D enhances innovation performance (through new or improved products or faster introduction of such new and improved products) and firm productivity (through the reduction of innovation costs and/or the utilization of economies of scale, scope, or learning) (Arvantitis, 2009). In concrete terms, there are four major firm capability areas which can be provided by open innovation activities resulting in knowledge transfer, as seen in Figure 6 (Tran, Hsuan, & Mahnke, 2011):

Figure 6: Open Innovation's Benefits to Capabilities

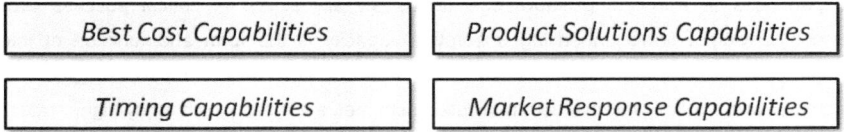

Source : Own Illustration based on (Tran, Hsuan, & Mahnke, 2011)

- Best cost capabilities: the key value-added dimension here is the lowering of costs across the board for all innovation activities.
- Product solution capabilities: the key-value added dimension here is the offering of new and enhancing of current product attributes.
- Timing capabilities: the key value-added dimension here is the shortened time to market for a product seen as a result of cooperation efforts.
- Market response capabilities: the key value-added dimension here is the reduction of hit or miss risks and the ability to generate more worth for customers.

The ability to create and import knowledge through systems of innovations is considered critical not only to the success of firms, but to regions and countries as well. Research centers and academic institutions are widely regarded as key players to regional economic development and form an irreplaceable part of the knowledge economy. Common models of university to industry partnerships help to support regional innovation systems. Specifically, this translates into benefits seen to individual firms, but also to the regional economy. Research frontiers can be driven forward by proactive and regular use of such institutions, resulting in the creation of knowledge hubs within regions. Ultimately, these hubs can be made available for less active firms in advancing their own innovation, strategic, and operational needs (Oakes, 2010).

Universities and other academic institutions are regarded as the backbone of any regional innovation system. Through attracting human capital and stimulation of entrepreneurial talent, they are often found in areas of high innovation activity. Other positive externalities emanate from these institutions as well, such as human capital spillovers, research lab collaboration, and spin-off creation (Graf, 2006).

Success Factors

A crucial aspect to advancing and promoting knowledge transfer is the capability of a company to understand and recognize the value of external information sources, while at the same time being able to assimilate and apply it to their own existing knowledge set. As a result, the concept of absorptive capacity takes a central role in understanding successful incoming spillovers. The ability to use and judge this new knowledge is a function of prior knowledge assimilations and generally succeeds best when firms have made similar experiences in transferring knowledge (Moos, et al., 2011). A critical success factor accompanying a firm's underlying absorptive capacity is the form and function of inter-organizational communication. It can be assumed that firms which have developed more complex and dynamic internal communication schemes are those which enjoy higher rates of absorptive capacity. The linkages within a firm are also paramount in aiding the assimilation of new knowledge - how the firm itself is structured in terms of hierarchal positions and reporting lines (Spithoven, Clarysse, & Knockaert, 2009).

There are several other concrete mechanisms which firms may utilize to increase absorptive capacity. By allocating resources differently, firms can modify their patterns by (Agrawal, 2001):

- recruiting graduate students who can both bring in new incoming spillovers but also raise the overall ability of the firm to learn.
- hiring professors as consultants and force the firm to look at new knowledge in different ways.

- modifying internal incentives to innovate and thereby motivating employees to be more proactive in their learning processes.
- working with universities and other academic institutions.
- participating in research consortia which exposes the firm to like-minded organizations.
- sending company employees for further education which can help broaden their knowledge base but also prepares them to accept new ideas from different sources.

Innovation networks must be given serious consideration as a method to organize knowledge transfer and integrate new concepts into a firm. Such networks have been associated with extending the scope of how an organization learns and integrates, as well as increasing, promoting, and fostering organizational flexibility. This leads to an overall condition where a company is better prepared to understand where to find new sources of knowledge, how to recognize those sources, and ultimately which mechanisms should be used to bring that knowledge into the confines of the firm (Liebeskind, Oliver, Zucker, & Brewer, 1996).

Transferring knowledge from one source to another operates more fluidly when the source and the recipient share some type of common knowledge. This common knowledge can take on multiple forms, such as a related industry, training, educational background, basis technology, management style, innovation goal, or personal background issues. It is for this reason that sharing knowledge across cultural or regional lines can be a more difficult proposition due to greater expected differences in the knowledge partners (Reagans & McEvily, 2003). Overall, there are three general context areas which determine whether knowledge transfer between two parties will be accepted and successful (Inganäs, 2008), as shown on the next page in Figure 7:

Figure 7: Knowledge Transfer Context

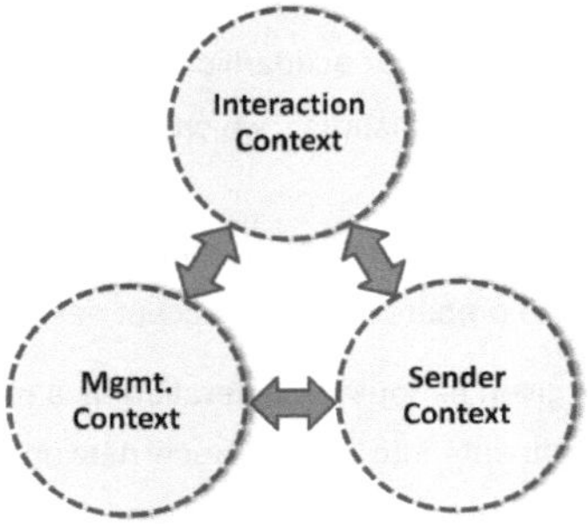

Interaction
Context

Mgmt.
Context

Sender
Context

Source : Own Illustration based on (Inganäs, 2008)

1. **Interaction Context:** This refers to the common knowledge concept and how the two knowledge partners must have some shared foundation to be able to initiate and sustain a profitable cooperation.

2. **Sender Context:** The original source of the knowledge must be able to provide accurate and appropriate information which is beneficial to the receiver.

3. **Management Context:** The approach taken by the management of the receiving party will determine how well the information is integrated and finally used. There is no correct approach, however the style must be consistent with the firm culture.

More implicit and complex forms of knowledge all have a geographic dimension which plays a critical role in determining success of transfer, as well as innovation processes. This geographical dimension can be positively influenced by policies and framework conditions (Reichert, 2006). A considerable amount of knowledge which can be considered complex and worthy of transmitting has an inherent sticky, or difficult, nature. The opportunity and actual costs of attempting to transfer such sticky knowledge rises as the distance increases between knowledge partners (Uotila, 2008). This points to regional systems of innovation being more successful than national or international ones. The difficulty in coordinating the different aims and knowledge interests of the actors in a regional system are less than in larger ones, while the base of knowledge is generally wider and broader than merely localized systems.

Problems Associated with Knowledge Transfer

As with any organizational process, the successful transfer and integration of knowledge can be impeded by several obstacles (Arvanitis, Kübli, & Wörter, 2005). Several competency deficiencies can contribute to blocking the process resulting in gaps in the essential competencies needed in order to undergo any knowledge process. A lack of information by both sender and receiver of knowledge transfer can be seen as a central impediment. This can refer to several aspects of the process, from no contact information in finding the correct partner to an underdeveloped internal skill set required to absorb knowledge. Deficiencies in the knowledge partner point to a missing appropriateness of the knowledge required. Unknown costs, risks, and uncertainties to the process have slowed companies motivations to begin their own process of knowledge transfer. Examined further, there can be a lack of firm financial resources for transfer activities - an unfortunate situation, as knowledge transfer itself often results in alleviating this very problem. Often, the R&D orientation of the knowledge partner is not interesting for a firm because the outcome cannot be commercialized. Institutional and organizational obstacles pose a problem when bureaucracy is considered to be too cumbersome to get involved. This perception is often associated with larger research projects overseen by governmental agencies and departments. Internal firm deficiencies also contribute to stopping the process. There can be a general perception within a firm that their research questions are not interesting enough to warrant the search for a knowledge partner.

Attention has been given to the overall reduced innovation/knowledge transfer activities of SMEs and why they appear to be lacking when compared to larger companies. Since innovation ability and knowledge transfer are so closely related, a firm's lack of innovation can be seen as a detriment to its information gathering abilities. Among these problems which SMEs face are (Ziltener, 2011):

- Management of SMEs have little or no time for innovation because they are too concerned with the operative activities of the firm.
- Innovation as a whole requires additional resources to be procured. This relates to the additional costs of infrastructure and/or dedicated personnel.
- SMEs lack the necessary network of contacts which is more commonly associated with and readily available for larger companies. SMEs rarely know what can be done for them by academic or research partners.
- Due to the inherit risks involved in investing in innovation, SMEs tend to shy away from radical projects and concern themselves with more incremental innovations. These are not the kind of projects which lead to real and profound change within the firm and/or industry.

During any knowledge transfer process, some problematic issues can and will be diagnosed easily and resolved relatively routinely by those involved in the transfer. However, not all problems will. The recognition and resolution of these issues could be out of the range of the firm's competency level and the resourcefulness of the actors involved. These kind of problems generally necessitate ad-hoc solutions in the form of additional and more personal interaction between the partners, non-standard skill development, allocation of different resources, and, where appropriate, involvement of higher hierarchal inter-organizational actors in transfer-related decisions. These actors, such as senior managers or consultants, would perhaps not have been usually required but changing circumstances force their involvement (Szulanski, 1999).

The OECD has come up with a list of generic lessons learned from several open innovation case studies aimed at preventing future problems arising from the practice of knowledge transfer. Among the most important are (OECD-3, 2008):

1. Open innovation requires a differentiated approach to knowledge sourcing and development: External cooperation needs to be coupled with internal diffusion strategies.
2. University knowledge is key in the exploration phase of any open innovation process: Collaboration between small firms and universities remains the best vehicle to undertake exploratory innovation projects.
3. A strategy about the use of intellectual property rights is key: Universities tend to be less savvy in this area, but it could lead to conflicts.
4. Trust is important: This can be coupled with a strong commitment needed by all involved.
5. There are organizational limits to open innovation: Trade-offs always exist between using different open approaches and requires a trial-and-error game plan on the part of the firms.
6. Creating a culture of open innovation in a firm requires rewarding teamwork and changes which promote internal and external collaboration: Work arrangements which encourage and reward risk taking is essential to overcoming barriers.

2.3 Literature Gap

As illustrated in the previous sub-chapters, there is extensive theory dealing with innovation networks as well as knowledge transfer and open innovation as a whole, as expected. Motivations, benefits of such techniques, associated problems, and other relevant aspects have been covered to the point of saturation. However, there is a gap in the existing literature concerning the project region of interest to KARIM - specifically, North-West Europe. The particular type and style of innovation networks found within this project region has not been catalogued to date. Moreover, the ad-hoc and informal nature of the observed innovation networks in this project region create a framework of data which is difficult to lay upon the foundation of the existing formal innovation network theory.

3 Methodology

The selection of a suitable methodology to accompany any research is an essential component which heavily influences and steers the undertaking towards the ultimate outcome. The methods chosen for this Master Thesis were based on linkages to theory and practical experiences. Therefore, the following chapter will provide an overview to how the empirical research process was conducted. The following will be explored in detail:

1. The research design created to be applied to the research.

2. The scientific approach used to solve the research problem.

3. Data collection issues, such as sampling considerations and the collection process.

4. An overview of the analysis methods chosen to break the data into manageable parts used to answer the research question and formulate implications.

An overview of the main methodological considerations is given below in Table 1.

Table 1: Methodology

Methodology	Method	Objective
Research Design	Multiple Case Study	Validation of theoretical propositions by focusing on a smaller data set and a contemporary problem
Research Approach	Inductive	Moving from specific cases to a general understanding of practical innovation networks
Research Strategy	Qualitative	Emphasis on process and richness of data, generalization in an analytical sense
Research Type	Exploratory, slightly Explanatory	Research gap closed, yet some relationships and patterns of network use identified
Data Collection	Secondary (from semi-structured interviews)	Interviews focus on joint construction of meaning
Data Analysis (Network)	Data Matrix/Cluster Analysis/Pattern Matching	Manipulation of Data to provide a tool in helping identify existence of innovation networks
Data Analysis (Firm)	Data Matrix/Frequency Analysis/Pattern Matching	Manipulation of Data to attempt to find similarities between firms using same network

Source : Own Illustration

3.1   Research Design

The methodology chosen to answer the research questions and validate the propositions was a Holistically Designed Multiple Case Study.

The case study design is an intensive analysis of an individual or multiple units, or cases, and the context, or subject. In this research, the cases are the multiple firms which were interviewed, while the context is their use of innovation networks. The primary advantage of using a case study method is that it is especially effective in answering "how/why" and some "what" research questions, it does not require control of behavioral events, and it focuses on contemporary problems (Yin, 2009) (Baxter & Jack, 2008).

Due to the multiple firms which were interviewed in order to fully understand the phenomenon, a robust multiple case study was selected. A total number of 28 cases (see Chapter 3.3 for additional information) was used, making this a relatively rich and detailed case examination. The methodology can be considered holistic because there is one context being examined in the final report - the existing innovation networks.

The research design applied to this report consisted of three stages. The stages were defined at the onset of this research.

Stage 1 - Define and Design: The problem was identified and defined in terms of what KARIM researchers expected. Existing theory was examined and reviewed in order to establish proper Qualitative propositions, along with two central research questions which existed in the gaps not to be found in literature.

Stage 2 - Prepare, Collect, and Initial Analysis: The sample frame, data collection techniques, and methods employed were defined and prepared by KARIM researchers. The data collection was also undertaken by KARIM. This stage was largely not addressed within this report, making the data secondary in nature.

Stage 3 - Analysis and Implications: Through various analysis methods, the collected data was examined to determine relevance in answering the research questions, as well as for future usable implications for KARIM and other key innovation actors. At the completion of this stage, the final report was written.

The design of the case study methodology for the final report can be seen on the following page in Figure 8.

Figure 8: Research Design

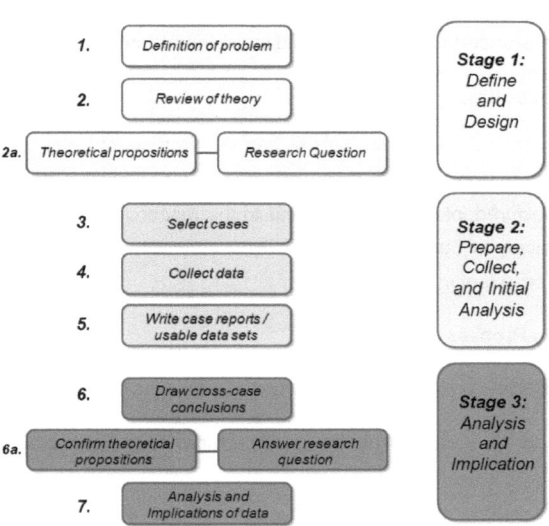

Source : Own Illustration based on Yin (Yin, 2009)

3.2 Approach

The case study is qualitative in nature. Of interest in the report is illumination and understanding as opposed to measuring casual determination in an empirical study. A general concept of innovation networks was defined through this process of discovery (Bryman & Bell, 2007). Due to qualitative research's inherit longer, more flexible relationship with respondents, this approach allows for a better understand of the inner workings of these networks. Moreover, the general problem was approached through an inductive, or bottom-up method. While there is an abundance of theory in this field and it was used to formulate propositions, the basic assumption was that there is no existing theory pertaining to specific innovation networks within the scope of this project.

Fundamentally, in social research there are three categories of research purposes - exploratory, explanatory, and description. These three categories differ in several aspects including, but not limited to, the way a research question and hypothesis is formulated and the method in which data is collected. None of these three have to be used exclusively in research but can be overlapping, as there is not always a clear boundary between them (Saunders, Lewis, & Thornhill, 2009). Because a portion of this Master Thesis is to fill in an existing research gap, the research conducted was primarily exploratory, yet also exhibits aspects of an explanatory research. While exploratory research is fundamentally used in areas that have not yet been explored in previous research, explanatory studies analyze causes and relationships in an attempt to identify patterns related to the subject studied - certainly a process undertaken to understand these innovation networks.

3.3 Data Collection

Considering the central role which KARIM has in this research, the selected cases to be studied as well as the collection of data was undertaken by members of KARIM itself in the form of semi-structured interviews. According to the research design from Figure 8, KARIM provided the majority of Stage 2. Therefore, the data which was used for the full report can be considered secondary in nature.

3.3.1 Sampling Process

The data is a collection of 28 interviews conducted with firms fulfilling the required criteria:

- The firms must have been within the **geographical scope** of the KARIM project (North-West Europe).
- The firms must have been recently involved in **innovation projects** with the goal of producing a new product, service, process, or organizational structure.
- The firms must have been considered **SMEs**, or less than 250 employees.
- An attempt was be made to cluster firms together from the **same industry**, for example, selecting multiple firms from the IT/Software industry.

The common assumption taken before the onset of conducting these interview was that the firms have been and are currently active within existing innovation networks, but this was not a prerequisite for the sample.

3.3.2 Data Collection Methods

KARIM conducted qualitative semi-structured interviews with these firms. A case study method was selected to prepare the final report due to the expected richness and depth of the conducted interviews. KARIM provided the interviews recorded, transcribed, and translated (if the interview language was not English). Additionally, a central Data Set was created which listed the most important aspects from every interview. This Data Set and all available transcribed interviews are contained within the CD attached to this Master Thesis. In order to address validity concerns as well as consistency, a detailed interview guide was used by the interviewees which guided the process towards usable data for the final report. Not only did the interview guide help maintain a level of consistency among the multiple KARIM interviewers who led these interviews, but it also organized the interview into several sub-units which was useful in the analyzing process. These sub-units included: the firm and interview partner, the innovation process within the firm, internal and external key factors/actors, cooperation and partnership, responsible open innovation, and general current needs. For the full interview guide, see Appendix A5.

The interviews were conducted with members of these firms who were in a position to be able to describe the innovation efforts of their respective organizations. Therefore, the potential list of interview partner positions was diverse and included the positions of CEO, Director, Marketing and Sales Manager, R&D Heads and Technician, Product Manager, and Research Officer. The duration of the interviews varied somewhat, depending on the relevance of the interview scope to the interviewed firm, however, the average interview lasted approximately one hour and took place at their place of business with one KARIM interviewer.

The full list of KARIM interviewed firms is found on pages 29-31 in Table 2. The numbers on the left of the table which has been assigned to each firm is the number which corresponds to that particular firm in any future graphics contained within this report.

Table 2: KARIM Interviewed Firms

		Firm	Branch	Interview Partner
Switzerland (CH)	1	Confidential	Motor Technology	Confidential Interviewee
	2	Confidential	Software/Energy	Confidential Interviewee
	3	Confidential	Energy	Confidential Interviewee
	4	Confidential	Energy	Confidential Interviewee
	5	Confidential	Life Science/Medical	Confidential Interviewee
	6	Confidential	Research/Testing	Confidential Interviewee
	7	Confidential	Energy	Confidential Interviewee
	8	Confidential	Life Science/Medical	Confidential Interviewee
	9	Confidential	Transportation	Confidential Interviewee

		Firm	Branch	Interview Partner
Germany (D)	10	Confidential	IT/Software	Confidential Interviewee
	11	Confidential	Webshops	Confidential Interviewee
	12	Confidential	IT/Software	Confidential Interviewee
	13	Confidential	IT/Software	Confidential Interviewee
	14	Confidential	Life Science/Medical	Confidential Interviewee
	15	Confidential	IT/Software	Confidential Interviewee
	16	Confidential	Life Science/Medical	Confidential Interviewee

		Firm	Branch	Interview Partner
England (Eng)	17	Confidential	Laboratory Support	Confidential Interviewee
	18	Confidential	Consulting/ Energy	Confidential Interviewee
	19	Confidential	Consulting/ Chemical	Confidential Interviewee
	20	Confidential	Consulting/ Energy	Confidential Interviewee
	21	Confidential	Horticultural Structures	Confidential Interviewee

		Firm	Branch	Interview Partner
France (F)	**22**	*Confidential*	Energy	Confidential Interviewee
	23	*Confidential*	Vision Systems, Printing	Confidential Interviewee
	24	*Confidential*	Coatings/Nano-structures	Confidential Interviewee
	25	*Confidential*	Transportation	Confidential Interviewee
	26	*Confidential*	Consulting/ Energy	Confidential Interviewee

		Firm	Branch	Interview Partner
Ireland (IRL)	**27**	*Confidential*	Glass Solutions	Confidential Interviewee
	28	*Confidential*	Research/ Testing	Confidential Interviewee

Source : Own Illustration

3.4 Data Analysis

The main objective of any data analysis revolves around closing the research gap between the theoretical findings of the literature review and the actual collected data, while simultaneously providing a framework to answer any research questions.

3.4.1 Data Matrix

The data from KARIM's Data Set was examined according to a specific set of variables, or dimensions (for a full list of these variables, see Chapters 5.1 and 6.1). These variables were then entered for each firm, creating a master list of information from which further analysis was able to be performed. This master list is known as the Data Matrix, By itself, the Data Matrix simply provides another way to view the data provided by KARIM. A Data Matrix was completed for the network perspective (see Appendix A2) and for the firm perspective (see Appendix A3).

3.4.2 Cluster Analysis

The purpose of any cluster analysis is to locate clusters of objects so that the objects of one cluster are similar to each other, while the objects located in other clusters are not similar.

Clustering methods are generally considered to be exploratory in nature, because it is typically not necessary to specify the number of clusters or certain characteristics of a cluster. The only real requirement is the need to specify variables and cases which should be used. (Bacher, 2002). Ultimately, a cluster analysis method was chosen to answer Research Question 1 in order to be able to manipulate the data from the Data Matrix, thereby providing a clearer view of the existence of any innovation networks. This cluster analysis is detailed further in Chapter 5.2.

3.4.3 Frequency Analysis

Because the firm perspective did not require manipulation of the data which would result in a set of clusters, a simple frequency analysis was chosen to answer Research Question 2. The data from the Data Matrix, being nominal in nature, does not lend itself to median and mode analysis. Because of this, a frequency analysis, or a simple counting of the number of answers given for a particular variable, is the closest method to finding a variable middle point and was therefore implemented.

3.4.4 Pattern Matching

For any case study analysis, a pattern matching logic is a powerful tool to compare empirically based qualitative data variables to each other (Trochim, 1989). The Data Matrix used in laying a framework for the cluster analysis and frequency analysis consisted of a series of variables detailing how each firm uses its own particular innovation network as well as important variables describing the firm. The variables were arranged in such a manner that patterns were able to be identified amongst the different firms, thereby creating an effective categorization where firms could be seen to be using similar innovation network types. For the network perspective analysis, when cross-referenced with the previously completed cluster analysis, the patterns seen amongst the cases were identified as individual networks. For the firm perspective, the patterns helped to support the frequency analysis in determining how a typical firm in a specific innovation network could logically be described.

4   Research Process

In the following chapter, the research process is expanded upon. Issues which could undermine the credibility and value of this Master Thesis are explored, as well as various ethical and other considerations.

4.1   Reliability

A common concern with any qualitative research - especially case study designs - is the inherent reliability issues which undoubtedly arise. Reliability refers to the concept of replication ability and repeatability of results and is a concern and addressed through several methods.

The case study protocol, as described in Chapter 3.1, was followed according to Robert Yin's (Yin, 2009) accepted procedure. Mr. Yin is considered an expert in the use and defense of the case study method. Additionally, emphasis was placed on documenting all procedures taken in constructing the full report.

Measures were taken to attempt to reduce the interpretative nature of the data by selecting analysis methods which would produce similar results for different researchers. Low levels of precision in data analysis techniques usually lead to a form of interpretative discretion by the researcher - suggesting patterns exist based on loose data (Trochim, 1989). In order to combat this lack of precision, the cluster and frequency analysis methods were deployed to illustrate statistically where similarities in firm's innovation behavior lies, thereby creating a more defensible foundation upon which the pattern matching could be laid upon.

It must be noted that if the variables in the Data Matrix were changed, it would affect the cluster analysis, frequency analysis, and the qualitative pattern matching afterwards. The number of possible networks which would have arisen from this different data is essentially endless, and therefore cannot be illustrated. Therefore, although fairly obvious, it should be stated that any change in the data would result in different results.

In order to further address the concept of repeatability, theory can be brought into the analytical stage. Existing theory illustrates multiple similarities between what the collected data reflected and what previous researchers found in their studies. This can be seen in how and why certain variables were chosen to be used for answering both research questions. Essentially, the theoretical propositions created at the beginning of the research process concerning innovation networks and the firms which use them helped to ground and guide the analysis process.

While the data used in this Master Thesis was secondary in nature, it is important to note that the actual collection of the data was done keeping consistency and reliability in mind. A common data collection procedure was established in a previous KARIM project, and all interviewers were asked to adhere to this procedure. Additionally, an interview guide created by the KARIM project leaders guided and steered the direction of each semi-structured interview, thereby increasing overall reliability.

While the issue of reliability is difficult to be fully satisfied in qualitative research, concepts such as consistency, credibility, dependability, and transferability are the most important concepts present during the research process (Baxter & Jack, 2008). Neutrality is an important concept here as well, as analysis of any qualitative data often has elements of interpretation in it. For a listing of how reliability issues were generally addressed, see Table 3.

Table 3: Reliability

Source : Own Illustration

4.2   Validity

Validity determines whether the means of measurement are accurate and measure what they propose to do. In other words, the extent to which the data was collected is a satisfactory indication of what it was intended to measure. There are there tests of validity in any research. Construct validity refers to the ability to identify correct operational measures for the concepts which are being studied. Internal validity refers to being able to establish relationships based on data. External validity defines a domain to which a study's findings can be repeated with the same results - a concept similar in scope to reliability.

Construct validity of this Master Thesis was achieved through a multi-pronged effort. There are many ways to measure how firms carry out their innovation activities, however, it was decided from the onset of the KARIM project that semi-structured interviews would give the richest and most in-depth detailed evaluation of these practices. This form of data collection allows for a story telling dynamic to unfold between the interviewee and interviewer which tends to produce spontaneous results not always seen when conducting straightforward quantitative surveys. For the purpose of this Master Thesis, the data which KARIM collected was used. Primary sources of data could have been collected, but considering the richness of KARIM's data it was deemed to have been superfluous.

Additionally, multiple firms from several different stages of development, industry, and innovation savvy where interviewed, ensuring a wide spectrum of data distribution. The chain of evidence, in terms of reviewing the data from KARIM, was kept as open and small as possible. KARIM researchers worked on the data, which was then used for this project - there were no other middlemen or organizations which worked on the data in-between.

Internal validity, and the efforts to find relationships in the data, was central to the analysis stage of this research. While statistically not as solid as a quantitative analysis, the methods used for this Master Thesis helped to establish both patterns and relationships. The cluster and frequency analysis coupled with pattern matching was a powerful tool in accomplishing this.

Lastly, the concept of external validity, or replication ability of the data, was adequately handled and covered in this report in the previous sub-chapter (See Table 4).

Table 4: Validity

Construct validity	**Internal validity**	**External validity**
Identifying correct operational measures for concepts being studied	*Seeking to establish relationships based on data*	*Defining the domain to which a study's findings can be repeated with the same results*
Semi-Structured Interviews	*Cluster and Frequency Analysis / Data Matrix*	*See Reliability Table*
Use of Multiple Firms	*Pattern Matching*	
Chain of Evidence in Terms of Reviewing Data from KARIM		

Source : Own Illustration

4.3   General Limitations

In addition to the above mentioned potential problems with validity and reliability that often accompanies qualitative and case study research, certain limitations exist which may undermine the integrity and rigor of this Master Thesis. A typical critical point of any case study research is that it provides no category and/or theoretical saturation of the sample. However, the goal of this research is not a representative generalization through empirical methods, but rather an analytical generalization. This can be further expanded upon by noting that analytical generalization is more akin to theoretical elaboration, where the result of this research can add to the existing pool of theory on innovation networks by illustrating real-life examples of where these networks exist.

The case study methodology itself can be a difficult method to gain consensus as to the correct procedure, however the method chosen for this Master Thesis was done after consulting three different and respected sources in the field of business research (Baxter & Jack, 2008; Bryman & Bell, 2007; Yin, 2009). It could be argued that the cases selected for this research were too small in scope to provide an appropriate data framework for analysis, as the semi-structured interviews averaged about one hour. However, the richness and detailed nature of the data which was taken out of these interviews makes up for their lack of length, as the interview guide assured that each interview would be steered towards the most important and relevant themes.

Finally, a critical point could be made that the interviewees focused their interviews on only examples of the best innovation promotion which these firms received as opposed to all aid given pertaining to innovation. This may reduce the scope of certain research goals, nevertheless the target of this Master Thesis is not the identification and classification of all innovation aid, rather examples of where it succeeded and created the best firm value. For a full listing of the research limitations, see Table 5 on the following page.

Table 5: Limitations

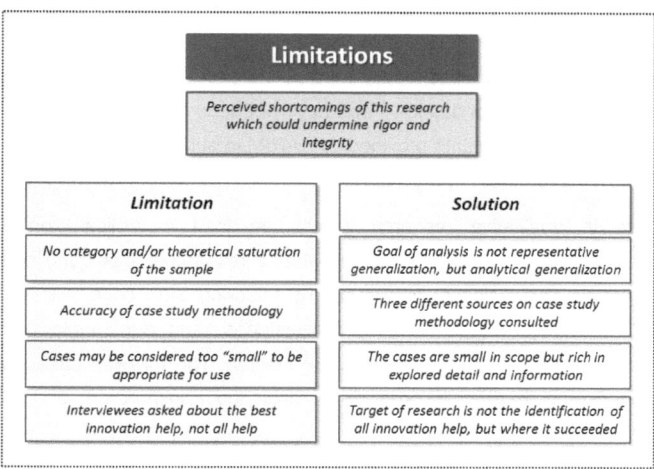

Source : Own Illustration

4.4   Delimitations

In terms of delimitations, this report does not make an attempt to determine which innovation networks are more effective and efficient, but rather how they are individually being used as per the research questions. Additionally, because the official scope of the project is North-West Europe, the KARIM data set focused primarily on five countries from the project region: Germany, France, Switzerland, Ireland, and England. This was essentially done for practical reasons, as the KARIM researchers used convenience sampling due to geographical concerns. Moreover, only SMEs were included in the research scope, which certainly provided a different set of answers had larger firms been also included. Additionally, specific measures to support these networks have not been established by this research. Their identification, classification, and similarities in firm structure was priority.

Finally, the presented innovation networks within this paper are all ad-hoc in nature, and as a result informal and, often times, spontaneous. Formal innovation networks are most certainly different in their scope and reach, but have been purposely left out of this report.

These limits on the research scope were done purposely to give the project a clear boundary, but further studies could be done to expand upon this research (see Chapter 7.4).

4.5   Ethical Considerations

As with any research, especially qualitative data collection through interviews, there are several ethical issues to be expected. Confidentiality and anonymity was offered to the participating firms of KARIM's interview process, however very few firms expressed a desire to have their information coded. The interviews were recorded and transcribed, and the majority of these recordings were used to validate information provided in the Data Set for this Master Thesis. All recordings and transcriptions provided by KARIM to support this project are located in a central databank and have not been distributed to third parties. After a period of two years, all metadata provided by KARIM to support this project will be deleted, so as to minimize the number of potential viewers.

Another issue in qualitative research which can lead to ethical considerations is transparency. It can be difficult to establish what a researcher has done or how he/she arrived at a certain conclusion of any study. By following Yin's case study protocol (see Chapter 4.1), the major steps, meetings, milestones, and documents have all been categorically filed and stored in a chronological fashion. The followed case study protocol can be provided upon request in order to increase the transparency, and ultimately the reliability and validity, of this research.

5   Results Perspective One : Network Perspective

The following chapter will answer Research Question 1, or from the perspective of the networks as a whole. The individual firm perspective which operate within these networks will be explored in Chapter 6 (Results Perspective Two).

5.1   Proposition One: Network Perspective

A powerful tool used to guide the analysis of case study data is the use of theoretical propositions. These are assumptions made about a certain topic based on existing literature and previous experience. The difference between a theoretical proposition and a hypothesis is a theoretical proposition is assumed to be true, while a hypothesis needs to be confirmed in order for it to be accepted.

Two propositions were formulated to guide this Master Thesis specifically by providing a foundation upon which the analysis of data could be laid. The first is explained below:

> *Proposition One : Network Perspective*. There are distinct groups of innovation networks within the project scope of KARIM. These networks can be categorized and described based on certain characteristics and patterns.

Existing theory has paid particular attention to being able to systematically categorize networks. The general problem is that there is no consensus to be found which can be applied to all types of innovation networks. Additionally, very little consensus exists concerning ad-hoc, or informal, networks. These networks are the dominant form found within the region of interest to KARIM.

Based on experience and a theoretical background, the networks found through the KARIM interview data analysis were categorized using 13 individual key variables, or dimensions. These fields and their descriptions are presented on the next page in Table 6. This completed set of variables for each firm forms the Data Matrix (see Appendix A2).

Table 6: Data Matrix Framework: Network Profile

Category	Description	Possible Codes
Size	The number of actors within a network	0,1,2,3,4,5,6,7,8
Geographic Range	Distance between the main actors	local, regional, national, international
Key Roles	The important actors within the network	customers, competitors, suppliers, consultants, universities, technical schools, govt. body, fairs, NGO, research, other firms, labs, public project, investor
Actors	Who the actors are	*list of innovation actors mentioned in interviews
Transaction Content Type	The nature of relationships between actors	information, goods, affect, power
Form of Innovation Supported	The type and scope of innovation projects	incremental/radical product, service, process, organization
Formality	The degree of formalization between actors	High formality, medium formality, low formality
Diversity	The diversity in the participants of the networks (age, industry, size, etc)	high diversity, mdium diversity, low diversity
Openness	Attitude of actors to being involved in other networks	open, closed
Intensity	The frequency of interactions between actors	strong intensity, weak intensity
Symmetry	The extent in which interactions are reciprocated - the balance of power	symmetrical, asymmetrical
Type of Cooperation	Activities done by exploitation of existing knowledge or creation of new	knowledge generation, knowledge exploitation
Financing	How has innovation been financed in the past	own capital, promotional money, investors

Source : Own Illustration

Size: This refers to the total number of actors which are present within a network. In most innovation network studies, this is heavily skewed by the researchers pre-conceived boundary as to what an actor is. This report understands actors as those defined below under the key roles dimension. In general, the more actors which participate within a network, the better exposure an organization has to unique knowledge. However, other dimensions within the network dictate how well that information can be transferred (Steward & Conway, 2009).

Geographic Range: The distance in which the majority of actors find themselves in relation to one another. For the purposes of this report, this can be sub-divided into 4 categories, adapted from (Ranga, 2009):

- Local: The innovation actors are located within a city or a metropolitan area, such as the Technopark in Zürich, the pharmaceutical concerns in Basel, or the technology clusters of Silicon Valley.
- Regional: All innovation activity is confined within a province, state, canton, or other recognizable regional area, such as North-Rhine Westphalia, Eastern Switzerland, or Northern England.
- National: The actors involved in the network are scattered throughout the political borders of a country, but are not generally found within the same regions of that country, such as Switzerland or France.
- International: At least one actor in the advancing of innovation activities is located across political country borders, such as Switzerland-France-Germany or England-France.

Key Roles: The actors which comprise a viable innovation network can be varied and diverse. The potential list of actors within the confines of the network include, but are not limited to (OECD-1, 1997):

- Internal firm sources: R&D research teams, project teams, marketing, finances, management.
- Market sources: Customers and clients, competitors, suppliers, distributors, consultants.
- Education and resource sources: Universities, technical institutes, government laboratories, fairs and exhibitions.
- Regulatory and governmental sources: government bodies and agencies, public projects, trade unions.
- Social and environmental sources: NGO's, pressure groups.

Transaction Content Type: The type of transaction in an innovation network is a reference to the flow between actors, underscoring the fact there is some type of exchange happening. There are 3 distinct types of transaction contents (Tichy, Tushman, & Fombrun, 1979):

- Affect: the exchange of friendship/contact information between actors. This refers to the classic definition of networking. The only real outcome of an affect transaction is the eventual contact with another actor.
- Information: the exchange of ideas, information, concepts, new directions, and know-how between actors.
- Goods: the exchange of goods, money, technology, infrastructure, human resources, or services between actors.

Form of Innovation: Certain networks are organized as such that they seemingly support one type of innovation better than others, based on the current firms using them and the typical output of these cooperation efforts. These forms of innovation can include incremental product, radical product, service, process, and organizational projects (Pittaway, Roberston, Munir, Denyer, & Neely, 2004).

Formality: Formality is a function of the level in which a relationship between actors within an innovation network is formally sanctioned. A formal relationship between the members of a network would generally be one governed by a contract or formal agreement. More informal networks represent linkages between actors that have emerged organically over a period of time (Steward & Conway, 2009). In general, the presence of governmental bodies and research institutions points to a more formalized network structure.

Diversity: The factor diversity is associated with not only the individuals operating within a network, but also the makeup of the organizations which use the network. In terms of individuals, diversity can refer to differences in individual's professions, ages, sex, or education. With regard to organizations within the network, the business sector, size, or type of organization can all vary between the members. High diversity in a network often points to a more probable presence of different and varied information and know-how, which usually raises the overall innovativeness of the network (Steward & Conway, 2009).

Openness: The term openness is a nod towards how readily available members of a network are to be connected with other actors in separate innovation networks. This underscores the fact that firms generally operate in multiple innovation networks simultaneously. Closed networks are comprised of a group of actors who interact primarily with each other only. In contrast, open networks are dominated by members who are linked to other groups or have no opposition to being linked to other groups (Steward & Conway, 2009).

Intensity: The intensity of any relationship within a network can be described as the frequency of interactions between actors within a certain time period. Generally, the intensity level of interactions between these actors is strong, middle, or weak in terms of its degree or power. Stronger intensity relationships within networks may lead to higher levels of interaction and change, but do not always correspond to the transfer of new and novel knowledge (Steward & Conway, 2009).

Symmetry: Symmetry refers to how balanced the power of transactions between members of an innovation network is. As a rule, a network can be considered to have an asymmetrical balance when transactions, power, and decision making are largely one-sided or dominated by a small amount of actors. A symmetrical network demonstrates a more democratic distribution on transactional relationship between its members. Symmetrical relationships are considered to be important in order to establish and maintain linkages for facilitating a freer flow of knowledge, but do not always result in a better quality of knowledge transfer (Steward & Conway, 2009).

Type of Cooperation: The multiple actors of any innovation network provide a range of services for all involved. Those services can include (Pittaway, Roberston, Munir, Denyer, & Neely, 2004):

- Knowledge generation: research, knowledge transfer and filling of competency gaps, partnerships, human capital.
- Knowledge exploitation: start-up support, coaching, consulting, training, benchmarking.

Financing: How innovation is financed within a firm structure depends on many factors and existing conditions. There are 4 main areas where a firm can procure the necessary financial resources deemed important to advance an innovation project: Own financing, or equity capital, investors such as business angels or venture capitalists, credit financing from bank institutions, and promotional money from government sponsored public innovation promotion funds.

5.2   Cluster Analysis

The Data Matrix created from the aforementioned variables was then further analyzed in order to determine where clusters of similar variables were. The cluster analysis was run using IBM's SPSS. The analysis was a hierarchical cluster, and the method chosen was median clustering using a squared Euclidean distance. This was done in order to create clusters that were as different as possible. A fixed number of cases, four, was decided after running a multiple case cluster analysis and examining the dendrogram, which showed a high relevance for exactly four distinct clusters. The variables used to run the cluster analysis are listed on the following page in Table 7.

Table 7: Cluster Analysis Variables

size0	size1	size2	size3	size4	size5	size6	size7	size8
0 Total Actors	1 Total Actors	2 Total Actors	3 Total Actors	4 Total Actors	5 Total Actors	6 Total Actors	7 Total Actors	8 Total Actors
region_local	region_reg	region_nat	region_int	trans_info	trans_good	trans_affect	rea_inst	govt_body
Local scope	Regional Scope	National Scope	Int. Scope	Transaction: Info	Transaction: Goods	Transaction: Affect	Research Inst.	Government
cust	firm	ngos	consultant	suppliers	distributors	fair_forum	investors	form_iprod
Customers	Other Firms	NGOs	Consultants	Suppliers	Distributors	Fairs/Forums	Investors	Increment. Product
form_rprod	form_serv	form_organi	formal	informal	high_diverse	low_diverse	open	closed
Radical Product	Service	Organizational	Formal	Informal	High Diversity	Low Diversity	Open	Closed
strong	weak	symmet	aysmmet	role_gen	role_exp	finan_own	finan_prom	finan_invest
Strong	Weak	Symmetrical	Aysmmetrical	Generation	Exploitation	Own Capital	Promotional	Investors

Source : Own Illustration

Every firm was given answers for the above-mentioned variables. Each variable was entered into SPSS as a positive and dummy variable. In other words, a common 1 and 0 system was chosen, with 1 representing a positive answer to the variable and 0 a negative. Upon running the cluster analysis with four pre-defined clusters, the following chart details how the cluster analysis placed each firm in the clusters (see Table 8).

Table 8: Cluster Membership and groups

Cluster	Cluster 1			Cluster 2		Cluster 3	Cluster 4
Firm Number	2,4,6,7,8,9,14,15,16,19,23,24,25,26,27			10,11,12,13,18,21,22		1,3,5,28	17,20
Identified Network	Knowledge and Learning	Resource and Finance Procurement	Public-Private Cooperation	Vertical Integration	Regional Clusters	International Scope	Isolated Islands

Source : Own Illustration

The identification of the four clusters and the firms which belong to each cluster can be seen as a stepping stone to the final step - the identification of networks. Since the data was already broken down into manageable parts, simple pattern matching techniques were used to compare the firms within a cluster in order to recognize any potential patterns which was present.

The result of this pattern matching was the identification of 7 networks - 3 networks from the first cluster, 2 from the second, and 1 from the last two.

5.3 Innovation Networks

The following is a list of the discovered networks based on the cluster analysis, pattern matching, theory, and experience. Seven networks have been identified:

1. Knowledge and Learning
2. Financial Procurement
3. Public-Private Cooperation
4. Vertical Integration
5. Regional Clusters
6. International Scope
7. Isolated Islands

These networks have been individually examined and analyzed. Presented here is a list of the important actors within each network, a typical network structure graphic, a profile of every network, and a short text description.

5.3.1 Network Type 1 : Knowledge and Learning

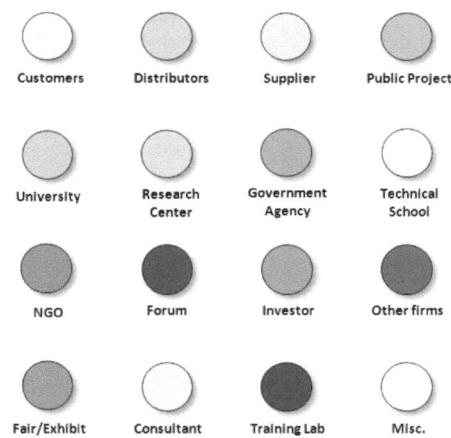

Network Type 1 : Knowledge and Learning

Important Innovation Actors within Network

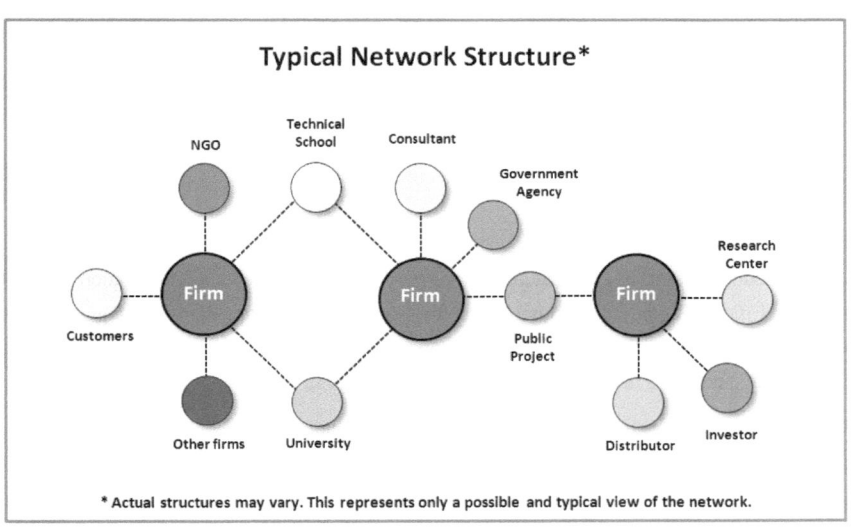

Typical Network Structure*

* Actual structures may vary. This represents only a possible and typical view of the network.

Table 9: Network 1 Profile - Network Perspective

Category		Network 1 : Knowledge and Learning	
Size (network)	Average number of actors in network	**More than 10**	Generally large, dynamic, and sprawling network structure
Size (number of firms)	Average number of firms in network	**4-5 average**	Multiple firms connected to one another through various innovation actors
Size (actors per firm)	Average number of actors per firm	**5-6 average**	Tends to be large
Geographic Range	Physical distance between actors	**National - International**	While international connections exist, most networks are nationally concentrated
Transaction Content	The nature of relationships for firms	**Information, Affect, Goods**	Information, new contacts, physical goods, new technology, finances, etc.
Form of Innovation	Type of supported innovation projects	**Incremental and Radical Product/Service, Org.**	All types generally supported by this network structure
Formality	Degree of formalization between actors	**Medium Formality**	Due to large number of actors, contracts and formal processes can vary
Diversity	Diversity (industry, size, scope of business) between actors	**High diversity**	As a result of the large number of actors and diverse scope of activities
Openness	Propensity of firms to be involved in other cooperation	**Open**	Generally firms involved in knowledge networks participate in other types
Intensity	The frequency of interaction between actors	**Strong**	Movement of transactions between actors and firms is generally frequent
Symmetry	The extent where power is balanced between actors	**Asymmetrical**	Large number of actors translates into some being more important than others
Type of Cooperation	Knowledge use: generation or exploitation	**Exploration and Exploitation**	Knowledge creation, exploitation, and exploration are equally used
Financing	How firms typically finance innovation	**Own, promotion, investors**	Because this structure is so large, all types of financing are seen

Source : Own Illustration

Description

Knowledge and Learning networks were the most prolific seen in the data collected and it can be assumed that they constitute a large percentage of the existing innovation networks through Europe. Once firms have committed to an open innovation strategy and seek external solutions to bridge internal knowledge and learning gaps, the array of potential actors and institutes available for use are large and diverse.

These networks tend to be large in size, both concerning the number of active actors and also the connections which bring different firms together. It cannot be said that there are important actors to this particular type of network, as all innovation actors which could potentially contribute to the advancement of a firm's innovation activities are considered critical to its proper functioning. One notable exception would be financial institutions or actors who are solely concerned with providing funding for innovation and not advancing knowledge.

Observed within the data set was a decidedly national feel to these networks – while several transaction types occurring in this network did indeed take place across existing country borders, in general the activities were bound by national lines. Presumably this is due to firms staying close to what they are familiar with – language, culture, and established lines of communications.

This type of general innovation network was seen to be conducive for all types of innovation support, including radical/incremental product development, service support, and internal organizational processes. Firms which participated in this varied network were given a wide range of benefits as a result, including, but not limited to, information, knowledge transfer, new contacts, physical goods as well as immaterial organizational support, and financial aid where needed. The network itself can be described as highly diverse and non-exclusionary, having little to no barriers for entry to participate due its multiple access points and heterogeneity of players. While each network is different and takes on its own individual character and flavor, it can be assumed that the relationship intensity between the actors within a Knowledge and Learning network are strong, underscoring the fact that there are a range of players and activities happening in order to support innovation.

This type of network takes on many shapes and sizes and can be considered the most general and non-specific identified network from KARIM's Data Set. Asymmetrical in nature, the actors located within have varying degrees of importance. While all firms which participate in this network are doing so with the intention of generating or exploiting knowledge, the possible ways to construct such a network are vast and, therefore, dynamic. It can be assumed that a veritable unlimited amount of possible network constructs exist today following the Knowledge and Learning mold.

5.3.2 Network Type 2 : Financial Procurement

Network Type 2 : Financial Procurement

Important Innovation Actors within Network

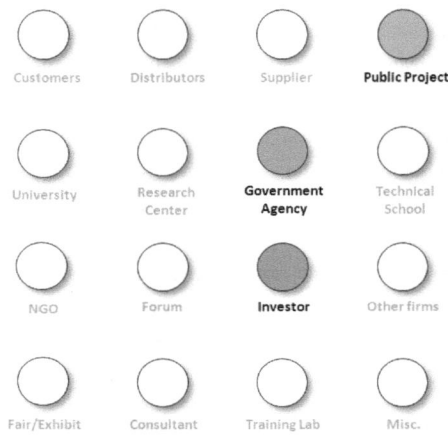

Typical Network Structure*

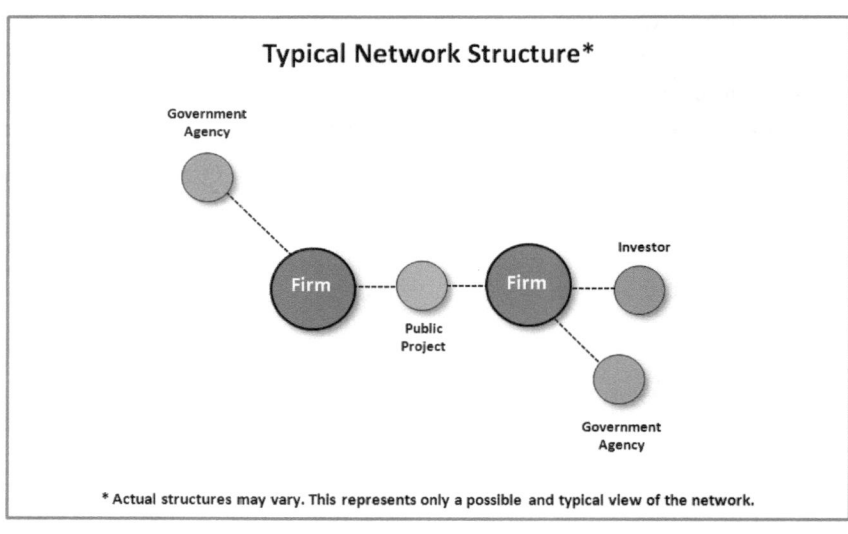

* Actual structures may vary. This represents only a possible and typical view of the network.

Table 10: Network 2 Profile - Network Perspective

Category		Network 2 : Financial Procurement	
Size (network)	Average number of actors in network	5-10	Smaller and more isolated networks concentrated only on finances
Size (number of firms)	Average number of firms in network	2 average	Connections between firms are not as common as in other structures
Size (actors per firm)	Average number of actors per firm	2 average	Each firm in this network averages about 2, usually an investor and/or a govt. agency
Geographic Range	Physical distance between actors	National - Regional	Except for EU public projects, firms stay national to procure finances
Transaction Content	The nature of relationships for firms	Goods	The basis of this network is the transaction of money.
Form of Innovation	Type of supported innovation projects	Products (necessary for radical)	Products are supported in this network, especially radical innovations
Formality	Degree of formalization between actors	High Formality	The transfer of money from one actor to another is generally highly formalized
Diversity	Diversity (industry, size, scope of business) between actors	Low Diversity	The diversity of the scope of activities of important financial actors is low
Openness	Propensity of firms to be involved in other cooperation	Open	Generally firms involved in finance networks participate in other types
Intensity	The frequency of interaction between actors	Weak	Movement of financial transactions between actors and firms is limited
Symmetry	The extent where power is balanced between actors	Symmetrical	Actors are generally traditional financial providers (govt., investors,)
Type of Cooperation	Knowledge use: generation or exploitation	Exploitation	Typically, knowledge is existing and needs to be exploited with financial resources
Financing	How firms typically finance innovation	Promotion, investors, credit	Several types of external financial sources (promotion, credit, investors, etc.)

Source : Own Illustration

Description

In contrast to some other types of networks, Financial Procurement networks exists to serve a specific purpose and function. The timely and proper procurement of financial resources has been identified in multiple interviews from the Data Set as a critical junction in a firm's innovation activities. It is also a leverage point with many innovation projects which often determines the success or failure prospects of any new undertaking. Several firms/networks were seen to specifically seek out the type of financial aid needed by especially young and upcoming enterprises.

This kind of network is quite smaller than the sprawling, unspecific nature of a Knowledge and Learning network. In general, firms usually have one to two partners which can satisfy the need to cover internal financial gaps. There is not a large and diverse field of actors for firms to choose from. Government agencies specific to countries often offer innovation promotion packages together with some kind of organizational and/or project support staff. Some universities and research centers have budgeted money for the specific purpose of doling out to advance innovation projects, but these can be considered governmental agencies, as the source of the finances are public coffers. Private investors and banks are an obvious alternative to using such public funds, but the drawback is that they are generally innovation network dead-ends and do not link the firm with any other innovation promoting institution or firm in the process.

There is a very high degree of formality associated with such networks, as financial resources are heavily regulated and tracked, meaning contractual and other bureaucratic issues for firms to deal with. Otherwise, the intensity between the actors in this network can be considered weak as the primary interaction revolves around the request and acquisition of finances. In a pure Financial Procurement network, information, knowledge, and other goods do not exchange hands. It is for this reason that this network can be considered a symmetrical construct, as there is no traditional role of power present aside from the actor giving out the funds. Nevertheless, it was observed that these kinds of networks do not exist in a vacuum and often are coupled with other kinds of innovation activities using the same actors. For example, it is rare that a public institution would make financial resources available for a firm and not be involved in some sort of knowledge transfer in order to ensure the success of the project.

Most firms in KARIM's Data Set expressed a clear desire to avoid the use of venture capital and other such bundled risk capital sources. While promotional money and private (preferably silent) investors were considered superior, a large portion of the interviewed firms did not use this Financial Procurement network, opting instead to attempt investing in innovation using their own equity capital.

5.3.3 Network Type 3 : Public-Private Cooperation

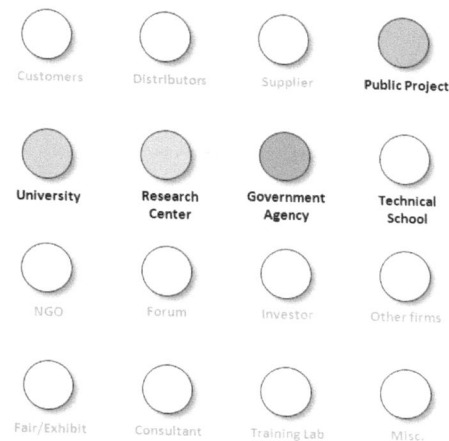

Important Innovation Actors within Network

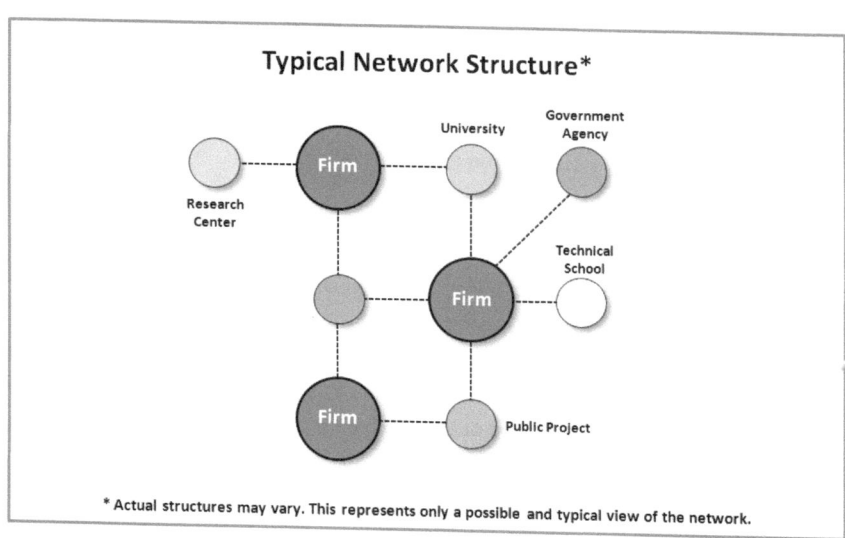

* Actual structures may vary. This represents only a possible and typical view of the network.

Table 11: Network 3 Profile - Network Perspective

Category		Network 3 : Public – Private Cooperation	
Size (network)	Average number of actors in network	**about 10**	These networks can be large in the size of actors involved
Size (number of firms)	Average number of firms in network	**3-4 average**	Firms connected by large public institutions
Size (actors per firm)	Average number of actors per firm	**3-4 average**	Due to the public nature of the actors, several firms at once can use one actor
Geographic Range	Physical distance between actors	**National - International**	Tend to be national in scope due to the governmental background of actors
Transaction Content	The nature of relationships for firms	**Information, affect, goods**	Information, contacts, physical and immaterial goods are transferred here
Form of Innovation	Type of supported innovation projects	**More product than service**	In general, products (radical and incremental) are supported in this network
Formality	Degree of formalization between actors	**High Formality**	Contact with governmental agencies/projects usually formalized
Diversity	Diversity (industry, size, scope of business) between actors	**Medium Diversity**	Universities and research centers together with governmental bodies
Openness	Propensity of firms to be involved in other cooperation	**Open**	Generally firms involved in public cooperation are also in other types
Intensity	The frequency of interaction between actors	**Strong**	Movement of transactions between actors and firms is frequent
Symmetry	The extent where power is balanced between actors	**Symmetrical**	Although somewhat different, actors are all from public sector
Type of Cooperation	Knowledge use, generation or exploitation	**Exploration and Exploitation**	Both knowledge exploration and exploitation are promoted in this structure
Financing	How firms typically finance innovation	**Promotion, own**	Often, cooperation is merely promotional money – not in all cases

Source : Own Illustration

Description

Throughout all observed networks in the KARIM project region, it can be said that no two other entities benefitted so much from each other's presence than the public sector coupled with private industry. This type of network, called a Public-Private Cooperation Network, plays upon the ever growing and important resource allocation set aside in public coffers for use by partners in the private industry. Theoretically, investing in the future innovation abilities of firms within a geographical area will increase the overall competiveness of the region and offset the cost to taxpayers in the long-term. This type of network was seen to be dominant in all regions of the project, but especially strong in France and Germany - two countries which have built a reputation on promoting innovation from within by using public resources.

In terms of size, these kinds of networks are not much smaller than the more common Knowledge and Learning network, owing to the fact that the majority of innovation actors are in some way connected to the public sector, either through financing, logistic support, or some other foundation. Similar to Knowledge networks, they tend to be national in scope due to the fact that large governmental bodies are the driving force behind them. However, the EU has several innovation actors which are currently being used by the member firms of KARIM's Data Set, which leads to a certain international aspect to this network as well.

The benefit which firms have using such a network are diverse and multi-faceted. Capital, knowledge, contact lists, organizational aid and support, coaching, and technology transfer are only a small portion of the reasons why interviewed firms decided upon using such systems. In this sense, knowledge is both created and exploited (when already existing) for a firm in using a Public-Private Cooperation network.

Government bodies and public projects are central actors in this construct. However, multiple universities, research centers, technical schools, and testing labs are supported in some fashion by their government, making them key actors as well. The formality level is generally higher here than a normal Knowledge and Learning network due to the fact that governments make an attempt to be transparent and fair about the allocation of their resources to the public sector. This leads to an inevitable strong-intensity relationship between firms and the governments which are supporting their innovation activities.

It has been observed that typically products (both incremental and radical) are the kind of innovation projects which are being supported by Public-Private Cooperation. It can only be speculated as to why services and/or organization innovations are underrepresented in this group, however, it may be a product of the small sample size.

5.3.4 Network Type 4 : Vertical Integration

Network Type 4 : Vertical Integration

Important Innovation Actors within Network

Typical Network Structure*

* Actual structures may vary. This represents only a possible and typical view of the network.

Table 12: Network 4 Profile - Network Perspective

Category		Network 4 : Vertical Integration	
Size (network)	Average number of actors in network	**Less than 5**	Firms usually isolated and not connected to other firms through actors
Size (number of firms)	Average number of firms in network	**1 average**	Due to very individual nature of actors within networks, firms not often connected
Size (actors per firm)	Average number of actors per firm	**1-2 average**	Customers, suppliers, and distributors make up the bulk of the actors
Geographic Range	Physical distance between actors	**National - Regional**	Majority of actors located in-country, with a high concentration being regional
Transaction Content	The nature of relationships for firms	**Information**	Information exploration is the main transaction type
Form of Innovation	Type of supported innovation projects	**Incremental Products / Services**	Products and services supported in this structure
Formality	Degree of formalization between actors	**Low Formality**	Often informal and personal contact between firm and actors
Diversity	Diversity (industry, size, scope of business) between actors	**Low Diversity**	While seemingly different, actors here are located along value chain
Openness	Propensity of firms to be involved in other cooperation	**Less Open**	Some firms involved in vertical integration only use this network structure
Intensity	The frequency of interaction between actors	**Weak**	Frequency of interaction traditionally not routine, but rather sporadic
Symmetry	The extent where power is balanced between actors	**Symmetrical**	All actors of relatively equal importance
Type of Cooperation	Knowledge use: generation or exploitation	**Exploration**	The aim is to create knowledge in such a network structure
Financing	How firms typically finance innovation	**Higher concentration of own**	While all forms of financing can be found, only own capital is seen often

Source : Own Illustration

Description

Often, firms look to more familiar avenues in order to support and advance their specific innovation projects. All firm undertake a series of activities in order to deliver a valuable product or service to the consumer markets, otherwise known as movements along the value chain. By coupling itself with partners along this value chain, a firm is said to be operating in a Vertical Integration network. The name vertical implies any movement along the value chain, whether it be forwards (distribution, sales, customers) or backwards (suppliers, material handlers). A true Vertical Integration network implies involvement from any of these actors located along the value chain. Often, this type of cooperation results in the identification and satisfaction of some kind of product/service need along the value chain which ultimately leads to an overall improvement in the worth perception held by customers at the end of the chain.

These informal networks traditionally are smaller and more isolated in nature than others, due to the very specific nature of each actor in relation to the firms which use them. For example, there are often multiple suppliers and distributors to a particular industry. It is fully possible that 10 firms from the same industry actively use 10 different suppliers. Moreover, customers often can be categorized into different market segments within the same industry. Generally speaking, these networks are found on a national level and rarely international.

It has been observed that a Vertical Integration network is more geared towards incremental products and service, as the main scope of transaction here focuses on the exchange of information only, while the relationship intensity is relatively weak between actors. Being fairly symmetrical based on the fact that all actors are located along the same value chain, it has been seen in the KARIM Data Set that several firms only used this type of network and eschewed others, underscoring the fact that this is a less open construct than others. On the other hand, it must also be said those firms which also were found to cooperate the most frequent and have the highest number of nnovation actors in their personal networks very often included customers input in their innovation processes, which would then be considered a rudimentary Vertical Integration network.

The inclusive of rival and competitive firms located within the same or a substitute industry would not be a part of this network structure, as that would represent a horizontal integration strategy, and are included in the more general Knowledge and Learning network category.

5.3.5 Network Type 5 : Regional Clusters

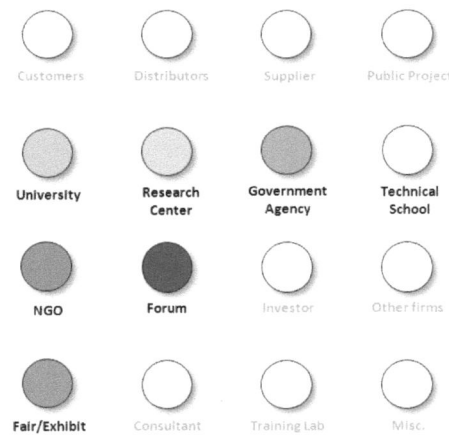

Network Type 5 : Regional Clusters

Important Innovation Actors within Network

Customers	Distributors	Supplier	Public Project
University	Research Center	Government Agency	Technical School
NGO	Forum	Investor	Other firms
Fair/Exhibit	Consultant	Training Lab	Misc.

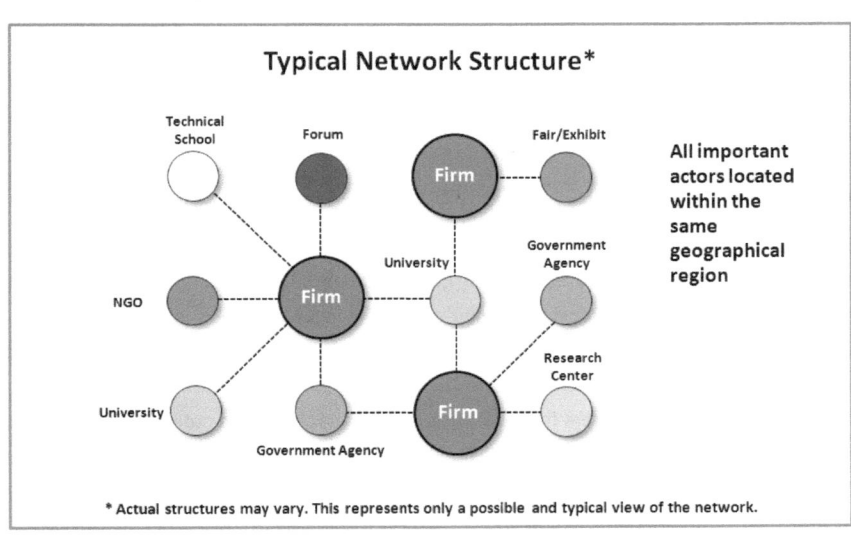

Typical Network Structure*

All important actors located within the same geographical region

* Actual structures may vary. This represents only a possible and typical view of the network.

Table 13: Network 5 Profile - Network Perspective

Category		Network 5 : Regional Clusters	
Size (network)	Average number of actors in network	**5-10**	Firms usually active with actors outside region, but 5-10 within
Size (number of firms)	Average number of firms in network	**3 average**	Connections happen in multiple places for firms
Size (actors per firm)	Average number of actors per firm	**3 average**	Diverse actors make up the range of interactions
Geographic Range	Physical distance between actors	**Regional**	By definition, the actors are all located within a geographical region
Transaction Content	The nature of relationships for firms	**Information, Affect, Goods**	Information, new contacts, physical goods, new technology, finances, etc.
Form of Innovation	Type of supported innovation projects	**Incremental and Radical Product/Service, Org.**	All types generally supported by this network structure
Formality	Degree of formalization between actors	**Medium Formality**	Some actors require more formality than others
Diversity	Diversity (industry, size, scope of business) between actors	**High Diversity**	Multiple actors from different sectors, industries, and scopes
Openness	Propensity of firms to be involved in other cooperation	**Open**	Generally, firms involved in regional clusters also use other types
Intensity	The frequency of interaction between actors	**Strong**	Movement of transactions between actors and firms is generally frequent
Symmetry	The extent where power is balanced between actors	**Asymmetrical**	As a result of the large number of actors, some are more important than others
Type of Cooperation	Knowledge use: generation or exploitation	**Exploration and Exploitation**	Knowledge can be explored and exploited in such a network
Financing	How firms typically finance innovation	**Own, promotion, investors**	All traditional forms of innovation financing are found here

Source : Own Illustration

Description

Regional innovation hubs have long been known to be particularly effective systems of innovation for firms, and the information gleaned from KARIM's Data Set supported this notion. The definition of a Regional Cluster network is loose and open to interpretation, as it may be difficult to properly identify a geographic region and receive consensus as to the credibility of that regional structure. However, wherever the defined borders of a region lie, it can be accepted that all actors must be located within this region in order for the network to be called a Regional Cluster. Six specific Regional Clusters were found in the Data Set: German Speaking Switzerland, English Lancaster University Link, Greater Dublin, Baden-Württemberg A and Baden-Württemberg B, and Central France/Paris.

The critical actors to such a Regional Cluster in theory are the same as the Knowledge and Learning network, as essentially it is the same network only on a regional scale. Customers, other firms, and investors have been purposely left out of this report's understanding of Regional Clusters due to the difficulty in establishing their geographic origin. Although both network types are similar in scope, Regional Clusters are smaller - less actors per firm, less firms per networks, and less actors in the entire network. These similarities include the underlying transaction of information, goods, contacts, financial, and physical resources. Additionally, the formality of such a network is medium, based on the diversity of actors, and the network itself can be seen as open, as participating firms frequently are involved in other types of networks.

This network structure supports all types of innovation, as a true Regional Cluster contains all relevant actors and channels to help promote a large range of innovations. It can be assumed that the very existence of the 6 discovered regions helps to attract even more firms to the area, resulting in more interest for innovation promoting actors. This creates a self-sustaining system which firms generally do not need to leave in order to fulfill their innovation requirements.

It is often the case that there are several firms using a Regional Cluster and are connected to each other through several actors, yet each firm is actively engaged in relationships with actors located outside the defined region. These actors would then not be considered part of the Regional Cluster, merely providing support aid or pieces of other networks which the firm finds itself it. It may also be the case that these firms are not purposely creating a Regional Cluster based on any reason other than it is comfortable and easier for them to work together with actors located nearby in the geographical region. Additionally, many firms have personal ties to universities and research centers which are located nearby. This university/school link often leads to other actors being brought into a particular firm's innovation processes.

5.3.6 Network Type 6 : International Scope

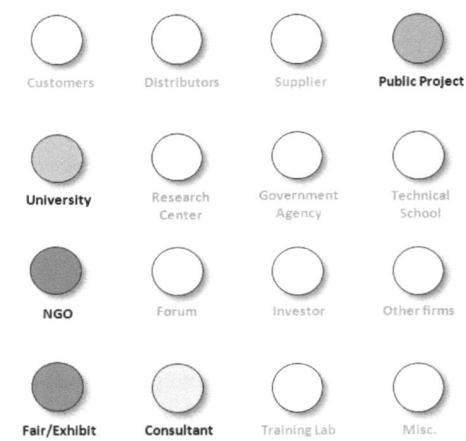

Network Type 6 : International Scope

Important Innovation Actors within Network

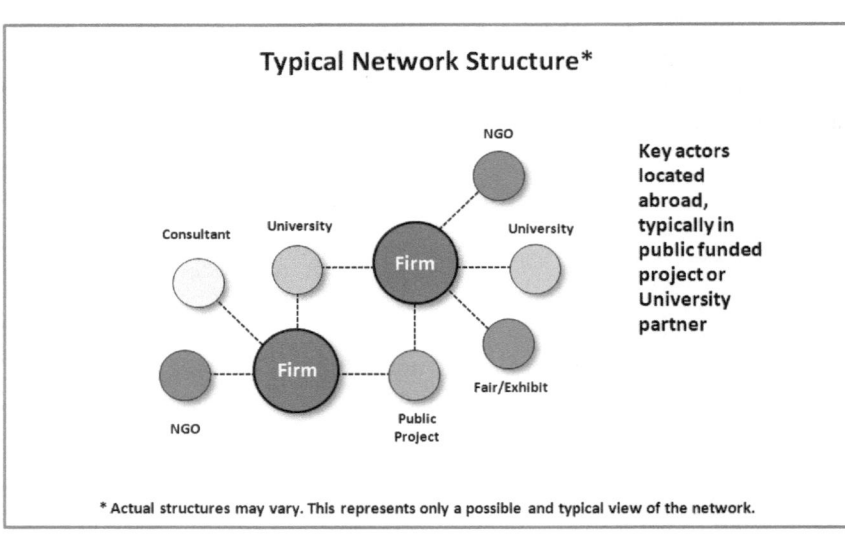

Typical Network Structure*

Key actors located abroad, typically in public funded project or University partner

* Actual structures may vary. This represents only a possible and typical view of the network.

Table 14: Network 6 Profile - Network Perspective

Category		Network 6 : International Scope	
Size (network)	Average number of actors in network	**Less than 5**	Smaller and more isolated actors compared to other networks
Size (number of firms)	Average number of firms in network	**2-3 average**	Firms usually connected by large EU public projects
Size (actors per firm)	Average number of actors per firm	**2 average**	International connections are seen more seldom, therefore there are less actors
Geographic Range	Physical distance between actors	**International**	By definition, the actors are all located in a different country from where the firm is
Transaction Content	The nature of relationships for firms	**Information, Affect, Goods**	Information, network information, and financial procurement from EU agencies
Form of Innovation	Type of supported innovation projects	**Product, very little Service, some Org.**	Generally, product innovations are supported the most here
Formality	Degree of formalization between actors	**High Formality**	Considering contact occurs often through EU channels, formality is high
Diversity	Diversity (industry, size, scope of business) between actors	**Low Diversity**	Actors here tend to be part of the EU governing or research arm
Openness	Propensity of firms to be involved in other cooperation	**Open**	Generally, firms involved in international scopes also use other types
Intensity	The frequency of interaction between actors	**Strong**	Movement of transactions between actors and firms is generally frequent
Symmetry	The extent where power is balanced between actors	**Symmetrical**	A smaller number of actors translates into a more balanced power structure
Type of Cooperation	Knowledge use generation or exploitation	**Exploration and Exploitation**	Knowledge can be created and exploited in such a network
Financing	How firms typically finance innovation	**Own, promotion**	Own capital and promotional money can be procured here, less international investors

Source : Own Illustration

Description

As expected, truly international cooperation is occurring within the project region. However, these border-free partnerships were observed in KARIM's Data Set less often than expected and generally in very isolated and fractured networks. These International Scope networks are defined as centering around a firm which uses one or more innovation promoting actors located in another country. Often, these international actors are large European Union projects, such as FP7. Due to their high potential in promoting, creating, and financing innovation, such large public projects generally attract other firms as well, forming the basis for a rudimentary International Scope network. There are other institutions which attract foreign firms and/or innovation actors as well, such as research centers, technical schools, and universities.

It has been seen that these networks are generally much smaller in size than others, averaging two or less firms and/or actors per network. Nevertheless, these types of relationships proved beneficial for the firms actively participating in them, as information, knowledge transfer, new contacts, physical goods, as well as immaterial organizational support and financial aid was received in some form.

It can be assumed that the average size of this International Scope network is kept small due to several reasons. Firms tend to be more wary in engaging in open innovation with culturally and geographically dissimilar institutions. It is for this reason that not only is the network structure itself small, but firms do not generally work with multiple international actors at the same time.

Interestingly, International Scope networks seemed to support incremental and radical product innovations the most within KARIM's Data Set, while service innovations where not heavily represented. In terms of formality, any dealing with the European Union must be considered rather formal - this is the perception of firms who use this network structure as well as those firms who purposely avoid it. Due to the small number of actors, it can be considered a low diverse network with symmetrical bonds between the players. Another aspect of using such a network structure is the access to European Union funding, which was a driver for many firms in participating.

This report's description of International Scope actors does not include customers, other firms, investors, or any other actor where it was not specifically mentioned that they are indeed operating on the international level.

5.3.7 Network Type 7 : Isolated Islands

Network Type 7 : Isolated Islands

Important Innovation Actors within Network

Customers	Distributors	Supplier	Public Project
University	Research Center	Government Agency	Technical School
NGO	Forum	Investor	Other firms
Fair/Exhibit	Consultant	Training Lab	Misc.

Typical Network Structure*

Firm

*Actual structures may vary. This represents only a possible and typical view of the network.

66

Table 15: Network 7 Profile - Network Perspective

Category		Network 7 : Isolated Islands	
Size (network)	Average number of actors in network	0-2	Innovation connections happen sporadically
Size (number of firms)	Average number of firms in network	1	Firms are not able to be connected to each other due to small amount of actors
Size (actors per firm)	Average number of actors per firm	1-2 average	Innovation generally happens in a closed environment with these firms
Geographic Range	Physical distance between actors	Local - Regional	When using external sources, actors are usually local
Transaction Content	The nature of relationships for firms	Information	Information gathering is the most important type of transaction
Form of Innovation	Type of supported innovation projects	Mainly Service, some Incremental Product	The majority of firms support a service innovation with this type of structure
Formality	Degree of formalization between actors	Low Formality	Interactions, when they happen, tend to be informal
Diversity	Diversity (industry, size, scope of business) between actors	Low Diversity	A small number of actors translates into a low diversity
Openness	Propensity of firms to be involved in other cooperation	Closed	Firms in this network often do not use any other type and remain closed
Intensity	The frequency of interaction between actors	Weak	Transactions between actors and firms is sporadic
Symmetry	The extent where power is balanced between actors	Symmetrical	A smaller number of actors translates into a more balanced power structure
Type of Cooperation	Knowledge use: generation or exploitation	Exploration	In general, new knowledge is not readily created but rather old info is expanded
Financing	How firms typically finance innovation	Own capital	Own capital financial sources is the preferred mechanism here.

Source : Own Illustration

Description

It was expected at the outset of this research that there would be firms which performed very little or even nothing in terms of open innovation and participation in innovation networks, and this was confirmed by the KARIM Data Set. Several firms had individually two or fewer innovation actors along with no discernible connection to other firms, thereby making them part of their own Isolated Islands network. These firms have chosen to execute a decidedly closed style of innovation for various reasons. Moreover, the firms which operate within this construct are generally not operating in any other type of network structure.

This Isolated Islands network was seen to support mainly services with some incremental products as well. The preferred mechanism for financing innovation within this system is an own equity capital, further underscoring the fact that firms who operate in this type of network do not engage in open activities, or when they do it most certainly is not for financial procurement.

It was observed that the sporadic open contact which does come from such an Isolated Islands network - however seldom it occurs - usually involves a form of vertical integration, with customers and/or supplies playing the main role. Weak and symmetrical, these networks can also be described as very informal.

This network type is indeed isolated in that any cooperation which is occurring between the firms and other actors are all almost certainly innovation dead-ends. That is, they do not lead to other connections or networks. The small amount of open innovation which is involved here is typically on a very local or regional scale. The assumption is that the regional impact of such cooperation is so low that it, at best, only marginally contributes to advancing the local and regional economy.

6   Results Perspective Two : Firm Perspective

The following chapter will answer Research Question 2, or from the perspective of the individual firms which operate within the existing innovation networks identified and described in Chapter 5 (Results Perspective One).

6.1   Proposition Two: Firm Perspective

The first proposition of this Master Thesis dealt specifically with innovation networks as a whole unit. This was an important step in identifying the different types and constellations which an innovation network can have. The second proposition assumes that these networks exist and examines specific dimensions, or variables, of firms who are using them.

> *Proposition Two : Firm Perspective.* Multiple similarities can be found in the firms using the same kind of innovation network, illustrating that certain types of innovation networks are better suited to certain kinds of firms.

The similarities in firm characteristic found between the individual companies using the same network are important leverage points which further help to define and understand each individual and particular network. From the perspective of the firm, it may be that all actors in the network are not being actively used. Firms are considered to be participating in a specific innovation network based on the type and scope of innovation actors which they use to promote their own internal innovation. Based on this, the data confirmed that firms use multiple innovation networks simultaneously.

Once again, based on experience and a theoretical background, the firms located within each innovation network were compared and contrasted using 9 individual dimensions. The specific dimensions used to analyze these firms were chosen based on their relevance to a firms general innovation activities. These dimensions and their description are presented on the following page in Table 16.

Table 16: Data Matrix Framework: Firm Profile

Category	Description	Possible Codes
Industry	Branch of Business	*too numerous
Stage	Stage of growth where the firm currently finds itself	Start-up, development, maturity, decline
Size	The number of employees in the firm	1-5, 6-10, 11-20, 21-50, 51+
Core Market	The geographical target market where the firm operates	Domestic, neighboring-country, Europe-wide, global
Business Model	How the firm generates value for its customers	Consulting, product-based, service-based, software, other
Value Chain Process	Portion of the firm's value chain which is affected by innovation activities	Marketing, R&D, production, distribution
Motivation to Participate	The motivation which led the firm to an open innovation strategy	Competency gaps, reduce costs, increase market share, finance/other resource, law
Timescale	The timeframe which a firm generally participates in open innovation	Project-based (short-term), competency need (medium-term), constant (long-term)
Innovation Formality	The degree to which the firm's internal innovation processes are formalized	formal, informal

Source : Own Illustration

Industry: The branch of business in which the firm is operating. The possible answers to this are numerous, although it can be expected that the most innovative firms are operating in a high-tech driven industry.

Stage: In general, businesses tend to grow and evolve through different stages which mirror how the actual industry itself is evolving (Mintzberg, Ahlstrand, & Lampel, 1998). Introduction phases (Start-up and Development) are generally fluid for firms and the industry because they are both fragmented in terms of knowing exactly what customers want and how to improve products. Uncertainty and experimentation are often seen by firms in these stages, until a specific product architecture can emerge. The difference between firms in the Start-up and Development phase can be blurry, but a generally accepted rule is that Start-ups have not established themselves or their products to any significant number of customers, while firms in the Development stage have earned some modest profits already. In the Development stage, more customers can understand the inherent value of the product. Innovation in both these early stages tends to be more radical in nature.

The Maturity phase of a firm/industry begins when growth slows and usually only the most efficient and well positioned firms survive in a relatively saturated market. Innovation seen here is generally incremental in nature. A firm/industry which finds itself in the Decline phase is being replaced by a disruptive and new technology which is eating away market share. (Christenson, 1997).

Size: The total number of employees in a firm. This is not necessarily an indication about how innovative that firm is. In low technology sectors, staff size is not seen to have any effect on innovation activities. However, in business fields with high technological opportunities, a positive correlation between firm size and innovative intensity can be seen (Archibugi, Evangelista, & Simonetti, 1995).

Core Market: The core market refers to the geographical markets where the firm targets its consumer segment. This can be categorized as either: Domestic, Neighboring-country markets (where internationalization has occurred but only initially to bordering countries), Europe-wide, and Global.

Business Model: Research-based firms (and university spin-offs based on research) typically can be categorized into four major areas of value creation for themselves and clients. These form the basis of a business model which creates the paradigm in which these firms operate. The four major types of business models are: Consulting, Product-based, Service-based, and Software. The link between business model and innovation efficiency is not clear, but it can be assumed that different business models form and are governed by different knowledge and technology transfer capabilities (Druilhe & Garnsey, 2004).

Value Chain Process: Refers to which area of the firm's value chain is affected by the innovation activities which it has undertaken. Traditionally, firms have only focused innovation activities on R&D, however increasingly areas such as marketing, production, and distribution can all benefit from cooperation (Trommsdorf & Steinhoff, 2007).

Motivation to Participate: Firms are individually motivated to open their innovation activities and participate in an innovation network. There are multiple reasons as to why a firm would begin this process. The most common reasons include: Competency gaps in the firm, pressure to reduce costs, chance to increase market share, financial/other resource procurement, and laws/regulations. The reasons which a firm have to pursue an open innovation strategy may directly correlate to which innovation network or networks it eventually chooses to join (Trommsdorf & Steinhoff, 2007).

Timescale: The involvement of a firm in an innovation network can be observed according to the timescale in which the cooperation occurs. Here, a firm can participate based on specific project needs (short-term), fulfillment of a competency need (medium-term), or a constant venture with an actor within a network (long-term). Short-term based cooperation generally revolves around procuring the necessary resources to see a particular project though. Middle-term generally involves a firm adding a new competency or skill to its stable of competencies. This may involve the firm attempting to branch into new areas/markets or use new technologies which previously were not available to them with the intention of adapting their value proposition. Long-term cooperation implies that the firm is constantly and dynamically engaged in the knowledge transfer process, usually with multiple partners and for multiple purposes. (Trommsdorf & Steinhoff, 2007).

Innovation Formality: Internal innovation processes and methods can be created in such a way that the firm is said to have either a formal or informal innovation process. While there is no clear boundary between the two, this report understands informal to mean little to no proper measures to capture, cultivate, and advance innovation in the firm (Trommsdorf & Steinhoff, 2007).

6.2   Frequency Analysis

The above mentioned variables were combined to create a Data Matrix for each network type (see Appendix 5). In each network, the answers given for the individual variables were counted and can be presented as a percentage of the total number of answers given in that particular variable category. Each network was then given an individual firm profile which lists the most frequently mentioned answer, along with the 2nd, 3rd, and 4th most common result.

6.3   Firm Profile

The following is a listing of the seven individual firm profiles created for each network. Additionally, each specific firm from KARIM's Data Set which participate in a network is listed, along with a text description. It must be noted that these are not profiles based on statistics in order to make sweeping generalizations or blanket statement about all other firms outside of KARIM's Data Set. Nevertheless, these profiles have roots in data and theory and can prove to be a good foundation point in making assumptions about commonalities of the firms using specific networks.

6.3 1 Firm Profile 1 : Knowledge and Learning

Table 17: Network 1 Profile - Firm Perspective

Network 1 Firm Profile : **Knowledge and Learning**	Typical Firm in Network 1	%	2nd Most Common	%	3rd Most Common	%	Least Common	%
Industry	Diverse	-						
Stage	Developing/Maturity	83%	Start-up	17%				
Size	6-10 employees and 21-50 employees	48%	11-20	17%	1-5	17%	51+	8%
Core Market	Global	39%	Europe-Wide	35%	Domestic	22%	Neighbor Countries	4%
Business Model	Product-Based	52%	Software	20%	Service	16%	Consulting	12%
Value Chain Process	R&D Processes	46%	Marketing	21%	Distribution	18%	Production	18%
Motivation to Participate	Competency Gap	42%	Finances	31%	Market Share	26%	Law	2%
Timescale	Competency (Med.-term) and Project (Short-term)	72%	Constant (Long-term)	28%				
Innovation Formality	Informal	60%	Formal	40%				

Network Type 1: Knowledge and Learning
The following firms from KARIM's Data Set were found to participate in this network

25 Firms = 89% of total 1

Source : Own Illustration

Description: Knowledge and Learning Firms

Due to Network Type 1's main purpose of knowledge and learning promotion being so general and diverse, little in terms of dominant dimensions were found in the data which could be assumed to be indicative of common traits held by the majority of firms. Nevertheless, some assumptions can be made and expected to be found by other firms using this network structure.

It is of little surprise that this network was found to be actively used by the largest portion of firms in the data set - almost 90% of KARIM's firms can be found in this structure. It is plausible that such a network structure would equally have such large membership when examining firms not included in KARIM's data. In general, these firms are somewhat older and larger than newly founded Start-up enterprises, as more than one-third of the firms have more than 21 employees. The vast majority of the firms using such networks can be expected to be active in international markets. More than 70% of all firms from the Data Set using this type of network structure were active either in global or European-oriented markets.

A large portion of these firms are most likely to have a product-based business. The data found supports this by demonstrating more than two-thirds of the firms having business models which are centered on either products or software. Service and consulting-oriented firms were not prominently featured. However, this could merely be a function of the fact that overall in the Data Set product-based business model firms were the majority. Gaps in competency and the search for financial means are anticipated drivers which motivate such firms to participate in a Knowledge and Learning environment.

Clear assumptions about the internal innovation formality cannot be made based on this data and/or theory found. It is likely that most firms which engage in such a network structure have a relatively informal internal innovation process, but this is most likely due to the fact that the majority of SMEs in the project region do not have clear innovation processes defined.

The typical project timescale which firms have in such a network probably runs from simple, project-based fulfillment of needs to constant and intense contact between actors. Interestingly, it seems that firms who use this type of network structure are more prone to shorter and medium-term relationships with their innovation partners. This likely mirrors the real world situation where firms are reluctant to engage in long-term, binding, and inflexible relationships whenever possible. This is again a result of the general nature of this network structure.

6.3 2   Firm Profile 2 : Financial Procurement

Table 18: Network 2 Profile - Firm Perspective

Network 2 Firm Profile : **Financial Procurement**	Typical Firm in Network 2	%	2nd Most Common	%	3rd Most Common	%	Least Common	%
Industry	Diverse	-						
Stage	Developing	44%	Maturity	38%	Start-up	18%		
Size	21-50 employees	31%	6-10	25%	11-20	19%	51+	6%
Core Market	Global/Europe-Wide	94%	Domestic	7%				
Business Model	Product-Based	41%	Service	24%	Software	24%	Consulting	12%
Value Chain Process	R&D Processes	45%	Marketing	24%	Distribution	18%	Production	14%
Motivation to Participate	Finances	40%	Competency Gap	33%	Market Share	26%	Law	2%
Timescale	Constant (Long-term)	36%	Project-Based (Short-term)	34%	Competency (Med.-term)	30%		
Innovation Formality	Formal	60%	Informal	40%				

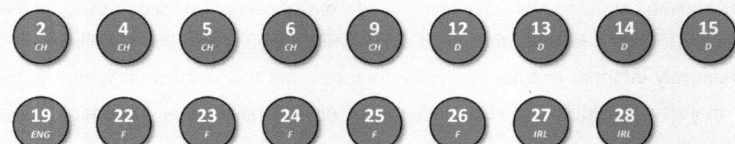

Network Type 2: *Financial Procurement*

The following firms from KARIM's Data Set were found to participate in this network

2 CH · 4 CH · 5 CH · 6 CH · 9 CH · 12 D · 13 D · 14 D · 15 D

19 ENG · 22 F · 23 F · 24 F · 25 F · 26 F · 27 IRL · 28 IRL

17 Firms = 61% of total **2**

Source : Own Illustration

Description: Financial Procurement Firms

The firms which were found to be active in the Financial Procurement networks possessed expected dimensions which one would anticipate to see in other firms which are actively searching to close financial gaps using open networks. These firms are more heavily in the development stage of their lifecycle - in fact, more than two-thirds of firms using this network from the Data Set were either Start-up or Development stage. It can be assumed that more mature firms either have their own means to finance innovation or have found other sources during the course of their development. The vast majority of these firms operate on a international stage, around 95%, which would be expected as they are attempting to procure large financial aid to support these campaigns.

This network possesses a smaller membership compared to the other innovation networks found within the Data Set. This is most certainly a function of most firm's reluctance to procure financial means outside the firm resulting in a, perceived or real, loss of total authority and control of an innovation project. This assumption is also confirmed through related statements given by the firms in KARIM's Data Set. Additionally, while the majority of firms seen using such a network are supporting a product-based business model, this is not as dominant as seen in firms using other types of networks. Service and software firms were seen to be strongly represented as well.

Interestingly, the data supports the theory that these firms tend to engage in more constant, or long-term innovation projects. This would make sense, as any financial commitment given to them from an innovation promotion agency (bank, government agency, research institution) would certainly be accompanied by contractual obligations and a stronger intensity of relationship between these actors. Moreover, these firms were seen to have more formal internal innovation processes as opposed to informal. This could feasibly be expected of other firms using a Financial Procurement network, as such processes may be a requirement for them to properly qualify for financial aid from public sources.

6.3.3   Firm Profile 3 : Public-Private Cooperation

Table 19: Network 3 Profile - Firm Perspective

Network 3 Firm Profile : **Public-Private Cooperation**	Typical Firm in Network 3	%	2nd Most Common	%	3rd Most Common	%	Least Common	%
Industry	Diverse	-						
Stage	Maturity	44%	Developing	39%	Start-up	17%		
Size	21-50 employees	30%	6-10	26%	1-5	18%	11-20	18%
Core Market	Europe-Wide	40%	Global	35%	Domestic	20%	Neighbor Countries	5%
Business Model	Product-Based	50%	Software	21%	Service	17%	Consulting	13%
Value Chain Process	R&D Processes	46%	Marketing	22%	Production	16%	Distribution	16%
Motivation to Participate	Competency Gap	40%	Finances	33%	Market Share	25%	Law	2%
Timescale	Competency (Medium-term)	38%	Project-Based (Short-term)	33%	Constant (Long-term)	29%		
Innovation Formality	Informal	58%	Formal	42%				

Network Type 3: Public – Private Cooperation

The following firms from KARIM's Data Set were found to participate in this network

1 CH   2 CH   3 CH   4 CH   5 CH   6 CH   8 CH   11 D   12 D

13 D   14 D   15 D   16 D   19 ENG   20 ENG   21 ENG   22 F   23 F

24 F   25 F   26 F   27 F   28 IRL

23 Firms = 82% of total 3

Source : Own Illustration

Description: Public-Private Cooperation Firms

The difference between firms which are active in Public-Private Cooperation networks and those in Knowledge and Learning networks is indeed minimal. This is due to the fact that public institutions play such a large role in any knowledge transfer-based network. It can be generally assumed that most firms which participate in Public-Private Cooperation networks are also part of a Knowledge and Learning network, however not vice-versa as Knowledge and Learning can entail other, more private aspects of KTT.

As a whole, Public-Private Cooperation networks may be more functional for firms which are slightly older and larger than those in Knowledge and Learning networks, as about half of the firms seen in KARIM's Data Set were mature and had more than 21 employees. There may also be a slight increase in the amount of formality found in the innovation capabilities of the firms which use such a network. This all has to do with the fact that dealing with external bureaucratic sources, while beneficial in the long-run, may prove difficult for smaller and less organized enterprises. Most transactions between actors of these kinds of networks are probably medium to long-term in nature, as any partnership with a government agency generally requires a time investment before the project can bear fruit.

Interestingly enough, the data showed a stronger tendency for firms who use this type of network to be more European-centered in terms of core markets, as opposed to purely global. It may be that these firms have become accustomed to European machinations and governmental regulations, and as such tend to innovate at this level as well.

As expected, both competency gaps and a lack of sufficient internal financial resources are the two main drivers which steered firms towards cooperation with a government body.

6.3.4 Firm Profile 4 : Vertical Integration

Table 20: Network 4 Profile - Firm Perspective

Network 4 Firm Profile : Vertical Integration

	Typical Firm in Network 4	%	2nd Most Common	%	3rd Most Common	%	Least Common	%
Industry	Diverse	-						
Stage	Developing	50%	Maturity	44%	Start-up	6%		
Size	21-50 employees	38%	11-20	25%	6-10	25%	1-5 and 51+	12%
Core Market	Domestic/Global	33%	Europe-Wide	26%	Domestic	20%	Neighbor Countries	7%
Business Model	Product-Based	47%	Software	24%	Consulting	18%	Service	12%
Value Chain Process	R&D Processes	41%	Distribution	22%	Marketing	22%	Production	16%
Motivation to Participate	Competency Gap	41%	Finances	30%	Market Share	27%	Law	3%
Timescale	Project-Based (Short-term)	48%	Competency (Med.-term)	29%	Constant (Long-term)	23%		
Innovation Formality	Informal	65%	Formal	35%				

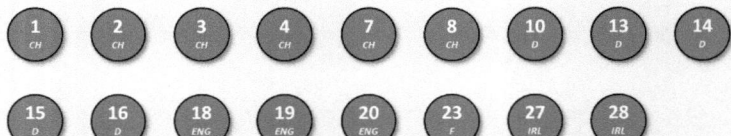

Network Type 4: Vertical Integration

The following firms from KARIM's Data Set were found to participate in this network

1 CH	2 CH	3 CH	4 CH	7 CH	8 CH	10 D	13 D	14 D

15 D	16 D	18 ENG	19 ENG	20 ENG	23 F	27 IRL	28 IRL

17 Firms = 61% of total **4**

Source : Own Illustration

Description: Vertical Integration Firms

Vertical Integration networks illustrate a somewhat different firm as compared to those firms who use other, more general networks. A good portion of the firms from KARIM's data which are located in this kind of network are not active in many other networks. This points to a special kind of firm which looks to integrate aspects of its value chain into innovation processes. Firstly, these firms tend to be younger - the majority of firms found were still in the developing stages. Additionally, these companies showed a propensity to being oriented heavily towards the domestic consumer market - in fact, less than half of the observed firms were internationally active. While the firms seen from the Data Set were equally active on the global stage, this dominant domestic role may stem from the fact that contact with suppliers, distributors, handlers, and customers is done the most effectively and efficiently when it happens in the home market.

Another strong feature of this kind of firm is its expected gravitation towards project-oriented, or short-term, timescales. This is to be anticipated, because it cannot be expected that contact with these members of the value chain would produce radical innovative results. Short-term innovation projects are primarily those concerned with incremental advancements. Long-term considerations are kept to a minimum for these kinds of firms.

As to be expected, competency gaps are the biggest drivers of motivations for firms to be in a Vertical Integration network, but concerns over distribution was seen as the second biggest driver for these firms. Considering distribution is a large feature of any value chain, this would seem logical.

Moreover, firms active in this network may show a high level of informality compared to other networks - fully two-thirds of all of the firms in KARIM's Data Set had very informal internal innovation processes. Clear and defined innovation goals generally leads a firm to gravitate to using other available innovation promotion actors, but an informal atmosphere within a firm may produce only attempts to include actors which are already in some way associated with the company, as is the case with all Vertical Integration actors.

6.3.5   Firm Profile 5 : Regional Clusters

Table 21: Network 5 Profile - Firm Perspective

Network 5 Firm Profile : **Regional Clusters**												
	Typical Firm in Network 5	%		2nd Most Common	%	3rd Most Common	%	Least Common	%
Industry	Diverse	-							
Stage	Developing	45%		Maturity	33%	Start-up	22%		
Size	6-10 employees	35%		1-5	24%	21-50	23%	11-20	18%
Core Market	Europe-Wide	38%		Global	31%	Domestic	25%	Neighbor Countries	6%
Business Model	Product-Based	47%		Software	26%	Consulting	16%	Service	11%
Value Chain Process	R&D Processes	44%		Marketing	23%	Distribution	21%	Production	12%
Motivation to Participate	Competency Gap	40%		Finances	33%	Market Share	25%	Law	2%
Timescale	Competency Need (Medium-term)	36%		Constant (Long-term)	32%	Project-Based (Short-term)	32%		
Innovation Formality	Informal	58%		Formal	42%				

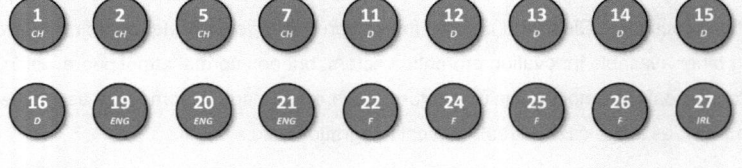

Network Type 5: Regional Clusters

The following firms from KARIM's Data Set were found to participate in this network

1 CH   2 CH   5 CH   7 CH   11 D   12 D   13 D   14 D   15 D

16 D   19 ENG   20 ENG   21 ENG   22 F   24 F   25 F   26 F   27 IRL

28 IRL

19 Firms = 68% of total **5**

Source : Own Illustration

Description: Regional Clusters Firms

There are multiple Regional Clusters operating in the project region of KARIM. Logically, it was observed that these firms have several identifiable dimensions which may predispose them to be better suited to working in a regional setting.

Seemingly, the firms within the Data Set which were identified to be staying within their own region in order to innovate were smaller and younger by comparison to firms which operate in and use other innovation network types. Fully two-thirds of the firms seen were either Start-ups or in the developing phase, while more than half had less than 10 employees in the firm. This underscores the fact that these firms probably understand the advantages to being active in an innovation network, but realistically don't have the resources to expand beyond the region. This is further supported by the fact that more than one-third of these firms use the network for financial support - a higher number than most innovation networks.

While the majority of these firms are also active in European based consumer markets, there is a high concentration of firms which are only active in domestic segments. This again suggests either a reluctance or lack of capability to leave the region. The fact that such a large number of firms use this type of network to support long-term innovation projects shows that there is indeed a high level of commitment on these firms parts which ultimately supports an innovation structure conducive to producing radical innovations.

These firms tend to show a low level of innovation formality, but this can be attributed to the fact that these firms are generally young and small. While R&D processes were the most important part of the value chain to be effected by these firms' participation in such networks, an even and fairly large showing for marketing and distribution details how these types of networks can be used for multiple needs which these younger firms have. It can be anticipated that other firms which use a Regional Cluster to promote innovation are similar in this aspect.

6.3.6   Firm Profile 6 : International Scope

Table 22: Network 6 Profile - Firm Perspective

Network 6 Firm Profile : **International Scope**											
	Typical Firm in Network 6	%		2nd Most Common	%	3rd Most Common	%	Least Common	%
Industry	Diverse	-							
Stage	Maturity	67%		Developing	22%	Start-up	11%		
Size	21-50 employees	38%		51+	25%	6-10	25%	11-20	12%
Core Market	Global	60%		Europe-Wide	40%				
Business Model	Product-Based	50%		Service	30%	Consulting	10%	Software	10%
Value Chain Process	R&D Processes	43%		Distribution	24%	Marketing	19%	Production	14%
Motivation to Participate	Competency Gap	38%		Finances	33%	Market Share	25%	Law	4%
Timescale	Competency Need (Medium-term)	50%		Project-Based (Short-term)	30%	Constant (Long-term)	20%		
Innovation Formality	Formal	56%		Informal	44%				

Network Type *6*: International Scope

The following firms from KARIM's Data Set were found to participate in this network

10 Firms = 36% of total *6*

Source : Own Illustration

Description: International Scope Firms

Compared to other innovation networks, the firm membership of the International Scope network is surprisingly low. Slightly more than one-third of the firms within KARIM's Data Set were engaged in some kind of innovation promotional activity with an actor or actors located over their own domestic borders. The overwhelming majority, about 67%, of these firms are currently in the maturity stage of their firm's life cycle, with a smaller portion being in the development stage. As one might expect, these firms were also large in terms of employees - more than half of the firms identified had 21 or more employees, and a quarter had more than 51.

Most of the firms seen to be using an International Scope network are using a business model which focuses on products and/or software, as service and consulting business structures play only a very small role here. This is perhaps to be expected, as tangible products or software may lend themselves to be more easily shipped across international borders as opposed to other revenue generating streams.

It can be anticipated that many of the firms which are using such a network would demonstrate a more formal approach to innovation, as structure, organization, and clarity of vision are all characteristics needed to be possessed by firms which attempt to be active in foreign markets. The data supports this by showing more than half of the firms in this network are indeed formal, a number much higher than the entire sample as a whole.

Next to general R&D promotion, distribution innovations tended to be somewhat important for the firms using this brand of network. This may be logical considering the firms are active in foreign markets and therefore are looking for unique and new ways to deliver their product to clients.

6.3.7 Firm Profile 7 : Isolated Islands

Table 23: Network 7 Profile - Firm Perspective

Network 7 Firm Profile : **Isolated Islands**										
	Typical Firm in Network 7	**%**	**2ⁿᵈ Most Common**	**%**	**3ʳᵈ Most Common**	**%**	**Least Common**	**%**		
Industry	Diverse	-								
Stage	Maturity	75%	Development	25%						
Size	1-5 employees	40%	51+	20%	6-10	20%	11-20	20%		
Core Market	Domestic	80%	Europe-Wide	20%						
Business Model	Consulting	40%	Service	20%	Software	20%	Product-Based	20%		
Value Chain Process	R&D Processes	43%	Production	20%						
Motivation to Participate	Competency Gap	80%	Finances	20%						
Timescale	Project-Based (Short-term)	100%								
Innovation Formality	Informal	100%								

Network Type *7*: Isolated Islands

The following firms from KARIM's Data Set were found to participate in this network

 10 D   17 ENG   18 ENG   20 ENG   21 ENG

5 Firms = 18% of total *7*

Source : Own Illustration

Description: Isolated Islands

While only 18% of the firms from KARIM's Data Set could be classified as firms which operate in the lonely Isolated Islands network (fewer than 2 innovation promoting actors per firm) it must be assumed that outside of the Data Set, this network type is larger. Membership is certainly based heavily on the industry type, as high-tech and newer industries tend to produce firms which understand inherently the need to innovate openly.

It is plausible that three types of firms are operating within this network structure. The first are firms which, as mentioned above, are in low-tech industries and do not place much importance on their own internal innovation capabilities. These were not found in the Data Set. The second are firms which are in classic high-tech fields, but have such dynamic and efficient internal innovation processes that the risks of external spillovers in open innovation projects is too high, therefore innovation is kept internal. These were also not found in the Data Set. The third type are firms which are older, domestic, and operate within high-tech industries. These firms have either not yet identified the advantages to innovation in a network or do not believe it applies to them. This was solidly seen in the data.

These firms tend to be older and, interestingly enough, the data showed that these firms were either very small or very large. However, due to the fact that membership to this network was so low, this information can easily be misinterpreted. In all likelihood, such firms are smaller in nature. They stay overwhelmingly only in domestic markets, with occasional jaunts into surrounding countries. Interestingly, it seems the business model of such firms would play a role here. Consulting firms were seen to be the most populous in the network, while all other business models were evenly distributed. It can be expected that consulting and service-oriented firms would feature prominently in such network structures outside of KARIM's Data Set.

Additionally, these firms do not have very fixed and formal innovation structures internally. When any kind of open innovation is attempted, it is done almost exclusively on a project-base level and not with the interest of creating long-term contacts to promote innovation in the future.

7 Implications and Conclusion

The final chapter of this Master Thesis will present a brief summary as to what was explored in the previous chapters, as well as the theoretical and practical implications which the results could potentially have. Finally, recommendations are given as to how this research could be expanded upon and continued by other researchers.

7.1 Summary

Innovation networks are increasingly becoming an important element in the innovation process for more and more firms. Considering that innovation networks have been linked to an overall more efficient project outcome and innovation itself is an important driver for economic and social success, it stands to reason that a detailed examination of said networks would be beneficial.

This Master Thesis focused on providing a detailed look at several innovation networks, based on qualitative firm data provided by KARIM (Knowledge Acceleration Responsible Innovation Meta Network), a European Union INTERREG IV B project. This research was structured and analyzed using a qualitative case study format, while KARIM's data was made usable by using a cluster analysis, frequency analysis, pattern matching, and existing theoretical foundations. Over the course of the research, seven types of innovation networks were found. These seven different types of innovation networks have multiple similarities, yet are fundamentally different from one another based on their purpose, use, structure, and firm membership.

Additionally, the seven individual innovation networks were further explored at the firm level, to be able to provide a comparison about the firms which are currently using them. Again, it was found that there are many similarities between firms which are using the same innovation network and a typical firm profile was established for each.

7.2 Theoretical Implications

One of the features of a qualitative case study approach, such as the one deployed in this research, is an inability to statistically generalize findings from a smaller sample to a larger population. However, due to the richness and complete nature of the data, one can begin to make analytical generalizations - grounded in theory and steeped in assumptions. Many analytical generalizations have already been made in the results portion of this Master Thesis. These generalizations apply to the individual innovation networks and firms found within. However, the concept of generalizing these findings can also be applied, in a larger sense, to the entire project region of KARIM.

It must be assumed that the seven innovation types found within this research are found with relatively high frequency within the KARIM project region. It can also be assumed, based on theory and fundamental knowledge, that there are other types of innovation networks which are different from the discovered seven, yet similar in their goal of providing a framework to aid firm's innovation capabilities.

Figure 9: KARIM Project Region and Networks

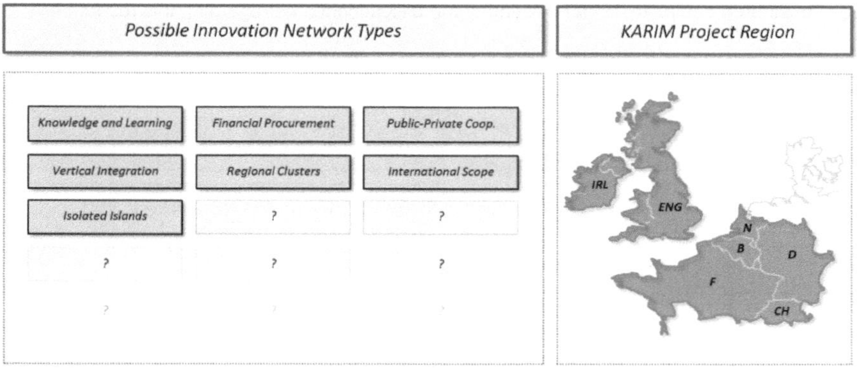

Source : Own Illustration based on (KARIM, 2012)

However, it is highly unlikely that new and previously unidentified innovation networks can be found which are more critical to firm's innovation success than the ones presented within this report. As a result, it is feasible to state that the aforementioned innovation networks from this Master Thesis represent the most important constructs of innovation actor/firm networks in the project region.

7.3   Practical Implications

While the number and type of additional innovation networks within the project region is difficult to determine, several practical implications of the seven discovered networks can be explored. The implications of any research imply the applicable use of the results, and this Master Thesis also delivers practical implications based on the results. By understanding the network structures and the firms which are prone to use them, different key players to the innovation process can address key concerns regarding innovation networks which they might have. This research conveniently separates these concerns into two levels. The Network level, or those concerns which are applicable to the networks themselves, can be addressed by examining the Perspective 1 (network) results (Chapter 5). The Firm level, or those concerns which are applicable to the firms which are using these networks, can be addressed by examining the Perspective 2 (firm) results (Chapter 6).

The practical implications of this work lie in these concerns. Different innovation actors will use this paper for different means. The list below in Table 24 shows several of these concerns. The concerns of these actors is presented in the form of a question which these actors generally have and can be answered with the aid of this Master Thesis. It is important to note that for several of these concerns, this research answers fully and completely the concern of the actor, while for others it is merely a starting point upon which more research and/or arguments can and should be laid upon. It must also be noted that this list should not be considered complete, rather a listing of the critical concerns regarding this research.

Table 24: Implications Concerns

Network Concerns	
Actor	**Concern addressed by this Master Thesis**
All	What kind of innovation networks are there?
Firms	What are concrete examples of regional/national innovation systems?
Firms	Is my firm geared more for international or domestic networks?
Firms	What specific research partners can help my innovation activities?
Firms	What specific network fits my innovation profile?
Govt. Agency	Which networks are operating at a regional/national level?
Govt. Agency	What policy can be implemented to help support these networks?
Govt. Agency	What can be done to reduce the amount of Isolated Islands networks?
Investors	What kind of networks have a current need for more private financial aid?
Schools and Research Inst.	How are some firms connected by a common School/Research Inst.?
Schools and Research Inst.	How far reaching can the cooperation with one firm go?
Schools and Research Inst.	What networks tend to demand more long-term and intense relationships?
Schools and Research Inst.	What can be done to attract more firm/actors to a Regional Cluster?
Schools and Research Inst.	What specific industry partners can work with us?

Firm Concerns

Actor	Concern addressed by this Master Thesis
All	What are the similarities in firms using the same innovation network?
Firms	How does my firm compare with other firms using the same network?
Firms	What are similar firms doing to promote their innovation in a network?
Firms	What specific industry partners can help my innovation activities?
Firms	What are some examples of firms using a certain kind of network?
Govt. Agency	Are there certain firms which should be targeted for certain networks?
Govt. Agency	How can some firms which are not using networks now be targeted?
Govt. Agency	Why do some firms use an Isolated Island network and can this be reduced?
Investors	What kind of firms turn to networks for financial needs?
Supplier/ Distributor firms	Are there certain aspects to firms which use Vertical Integration networks?
Schools and Research Inst.	Which innovation networks seems to best support spin-offs?
Schools and Research Inst.	What type of firm tends to use universities/research Institutes?
Schools and Research Inst.	What kind of firm can be targeted to use more public research?

Source : Own Illustration

The above illustrated table cannot be considered complete, rather a listing of the critical concerns which various key players related to innovation networks may have. By combining both theoretical and practical implications, this Master Thesis can have value for a diverse range of actors in providing a basis on multiple issues. It for this reason that the results of this research were left purposely general and broad, further supporting the fact that the implications here are wide-reaching.

7.4 Further Research Recommendations

This research has covered a broad and extended base of knowledge by presenting the innovation networks found within KARIM's project region and the firms which are using them. However, it can be expanded upon by using it as a foundation upon which newer and more targeted researches can be built. The following is a list of research questions which could potentially be answered by future endeavors.

1. What are some of the policy instruments which accompany and support the individual innovation networks?
2. What are the critical success factors which lead to these networks efficiently and effectively supporting innovation?
3. How are the individual actors located within each innovation connected to one another and what is the nature of their relationship?
4. Which innovation networks are the most effective in terms of providing an external innovation support system for firms?
5. What are the differences between how SMEs and MNCs are using innovation networks?

Additionally, a quantitative study could be done to accompany this research. This Master Thesis focuses on an inductive, general-to-specific model of research. The results can be used as a starting point to make assumptions about other firms and innovation networks not included within. However, a quantitative study could take a deductive approach in examining a large sample of firms, thereby giving credence to any statistical generalization which arises from such an undertaking.

91

List of References

Agrawal, A. (2001). University to industry knowledge transfer: Literature review and unanswered questions. *International Journal of Management Reviews, Vol. 3, No. 4* , S. 285-302.

Ahlstrom-Söderling, R. (2003). SME Strategic business networs seen as learning organizations. *Journal of Small Business and Enterprise Development, Vol. 10, No. 4* , S. 444-454.

Archibugi, D., Evangelista, R., & Simonetti, R. (1995). Concentration, firm size, and innovation: evidence from innovation costs. *Technovation, Vol. 15, No. 3* , S. 153-164.

Arvanitis, S., Kübli, U., & Wörter, M. (2005). Determinants of knowledge and technology transfer activities between firms and science institutions in Switzerland. *KOF Working Papers, ETH, No. 116* , S. 1-35.

Arvantitis, S. (2009). *How do different motives for R&D cooperation affect firm performance? An analysis based on Swiss micro data.* Zürich, Switzerland: KOF Working Paper, ETH, No. 233.

Audretsch, D., & Feldman, M. (1996). R&D spillovers and the geography of innovation and production. *American Economic Review, Vol. 86, No. 3* , S. 630-640.

Bacher, J. (2002). *Cluster analysis.* Nuremberg: University Erlangen-Nuremberg.

Barney, J., Wright, M., & Ketchen, D. (2001). Firm resources and sustained competitive advantages: Ten years after 1991. *Journal of management, Vol. 27, No. 6* , S. 625-641.

Bau, F. (2011). *KARIM : Projektübersicht (presentation).* HTW Chur, Switzerland: KARIM Project, Swiss partner.

Baxter, P., & Jack, S. (2008). Qualitative case study methodology: Study design and implementation for nove researchers. *The Qualitative Report, Vol. 13, No., 4* , S. 544-559.

Becker, K., & Bau, F. (2011). *Action 1: Interactive innovation map handbook for data collection.* KARIM.

Bolli, T., & Wörter, M. (2011, Februar). Competition and R&D Cooperation with Universities and Competitors.

Bryman, A., & Bell, E. (2007). *Business research methods.* New York, NY: Oxford University Press Inc.

Busquets, J. (2010). *Orchestrating Network Behavior for Innovation.* Copenhagen, Denmark: Programme for Informatics: Copenhagen Business School Department of Informatics.

Carayannis, E., & Campbell, D. (2006). *Knowledge creation, diffusion, and use in innovation networks and knowledge cluster.* Westport, Connecticut: Praeger.

Cassiman, B., & Veugelers, R. (2002). R&D cooperation and spillovers: Some empirical evidence from Belgium. *American Economic Review, Vol. 92, No. 4* , S. 1169-1184.

Ceglie, G., & Dini, M. (1999). *SME Cluster and network development in developing countries.* UNIDO.

Chesbrough, H. (2004). Open innovation: Renewing growth from industrial R&D. *10th Annual Innovation Convergence.* Minneapolis, USA.

Chesbrough, H. (1. March 2003). Open innovation: The new imperative for creating and profiting from technology. *Harvard Business School Press* .

Christenson, C. (1997). *The innovator's dilemma. When technologies cause great firms to fail.* Massachusetts, USA: The President and Fellows of Harvard College (Harvard Printing).

Conway, S., & Steward, F. (2009). *Managing and shaping innovation.* Oxford: Oxford University Press.

Conway, S., & Steward, F. (1998). Mapping innovation networks. *International Journal of Innovation Management, Vo.l 2, No. 2.* , S. 165-196.

Conway, S., & Steward, F. (1998). Mapping innovation networks. *International Journal of Innovation Management. Vo.l 2, No. 2* , S. 165-196.

Cooke, R., & Morgan, K. (1994). *The creative milieu: A regional perspective on innovation, in: Dodsson, M. Rothwell, The Handbook of Industrial Innovation.* Aldershot: Edward Elgar.

Cowan, R., Jonard, N., & Zimmermann, J. (2005). *Bilateral collaboration and emergent networks.* iNeck Thematic Network (5th Framework Programme of the European Commission).

Daniel, L., & Grigg, L. (2003). Inter-organizational networks, value creation, and the process of technology integration in R&D. *International Journal of Technology, Policy, and Management, Vol. 3, No. 1* .

Davenport, T., & Harris, J. (2007). *Competing on analytics - the new science of winning.* Harvard Business School Press.

de Man, A. (2008). *Knowledge management and innovation in networks.* Cheltenham, UK: Edward Elgar Publishing.

Druilhe, C., & Garnsey, E. (2004). Do academic spin-outs differ and does it matter? *The Journal of Technology Transfer, Vol. 29, No. 3* , S. 269-285.

Faems, D., & Van Looy, B. (2003). *The role of inter-organizational collaboration within innovation strategies: towards a portfolio approach.* Belgium: Vlerick Leuven Gent Working Paper, Ghent University.

Festel, G., & Boutellier, R. (2008). KMU-Finanzierung am Beispiel der industriellen Biotechnologie. *Innovation Management, No 9* , S. 99-99.

Freeman, L. (1991). Networks of innovators: A synthesis of research issue. *Research Policy Vol. 20, No. 5* , S. 459-514.

Gassman, O., Enkel, E., & Chesbrough, H. (2010). The future of open innovation. *R&D Management, Vol.8* , S. 213-218.

Gibbons, M., & Johnston, R. (1974). The roles of science in technological innovation. *Research Policy, Vol 3, No. 3.* , S. 220-242.

Gilseng, V., & Nooteboom, B. (2004). *Density and strength of ties in innovation networks: An analysis of multimeadia and biotechnology.* Eindhoven, the Netherlands: Eindhover Centre for Innovation Studies.

Gloor, P. (2006). *Swarm creativity: Competitive advantage through collaborative innovative networks.* New York: Oxfod University Press.

Gordon, I., & MacCann, P. (2000). Industrial clusters: Complexes, agglomeration, and/or social networks. *Urban Studies, Vol 37., No.* , S. 513-532.

Graf, H. (2006). *Networks in the innovation process: Local and regional interactions.* Cheltenham, UK: Edward Elgar.

Grant, R. (1996). Towards a knowledge-based theory of the firm. *Strategic Management Journal, Vol. 17* , S. 109-122.

Hautmäki, A. (2007). *Multi-channel innovation networks.* Helsinki: Theoritical Philosophy, University of Helsinki.

Hippel, E. V. (1994). Sticky information and the locus of problem solving: Implication for innovation. *Management Science, Vol. 40* , S. 429-439.

Hopkins, D. (2003). Understanding networks for innovation in policy and practice. *From: Networks of innovation, OECD Publications* .

Howard-Partners. (2007). *Study of the role of intermediaries in support of innovation.* Canberra, Australia: Department of Industry, Tourism, and Resources.

Inganäs, M. (2008). *Organizing and managing university-industry knowledge transfer - a study of the Swiss biotechnology sector.* Zürich, Switzerland: ETH PhD Dissertation Submission No. 177795.

INTERREG IVB North-West Europe. (n.d.). Retrieved July 10, 2011, from http://www.nweurope.eu/

KARIM. (2012). *New European Project 2011-2014 KARIM.* Paris, France: KARIM Region Innovation Center.

Kogut, B. (2000). The Network as Knowledge: Genertive rules and the emergence of structure. *Strategic Mangement Journal, Vol. 21, No. 3* , S. 405-425.

Küppers, G., & Pyka, A. (2002). *Innovation networks: Theory and practice.* Cheltenham, UK: Edward Elgar.

Liebeskind, J., Oliver, A., Zucker, L., & Brewer, M. (1996). Social networks, learning, and flexibility: Sourcing scientific knowledge in new biotechnology firms. *Organization Science, Vol. 7, No. 4* , S. 428-443.

Lord, M., & Ranft, A. (2000). Organizational learning about new international markets: Exploring the internal transfer of local market knowledge. *Journal of International Business Studies, Vol. 31, No. 4* , S. 573-589.

Marques, M., Alves, J., & Saur, I. (2005). *Creating and sustaining successful innovation networks.* Copenhagen, Denmark: Dynamics of Industry and Innovation: Organizations, Networks, and Systems.

McEvily, S., Das, S., & McCabe, K. (2000). Avoiding competence subsitution through knowledge sharing. *Academy of Management Review, Vol. 25, No. 2* , S. 294-311.

Meyer, M., & Utterbeck, J. (November 1995). Product development cycle time and commercial success. *IEEE Trans. Eng. Manage.* , S. 297-304.

Mintzberg, H. (1984). Power and Organization Life Cycles. *Academy of Management Review, Vol. 9, No. 2* , 207.

Mintzberg, H., Ahlstrand, B., & Lampel, J. (1998). *Strategy safari: A guided walkthrough the wilds of strategic management.* New York, USA: New York Free Press.

Moos, B., Beimborn, D., Wagner, H., & Weitzel, T. (2011). *Knowledge management systems, absorpative capacity, and innovation success.* ECIS 2011 Proceedings. Paper 145.

Narula, R. (2004). R&D Collaboration by SMEs: new opportunities and limitations in the face of globalization. *Technovation, Vol. 24, No.2* , S. 153-161.

Nooteboom, B. (1999). *Inter-firm Alliances. Analysis and Design.* London: Routledge.

Oakes, I. (2010). *The role of university - Business collaboration in influencing regional innovation, in: Innovation through knowledge transfer by Howlett, R.* Springer: Berlin.

OECD. (2011). *Policy Brief 2011: Regions and innovation policy.* Secretary-General of the OECD.

OECD-1. (1997). *National innovation systems.* Paris, France: OECD Publications.

OECD-2. (2005). *Guideline for collectin and interpreting innovation data - Oslo Manual.* Luxumbourg: OECD-EUROSTAT.

OECD-3. (2008). *Open innovation in global networks.* Paris France: OECD Publications.

Osterloh, M., & Frey, B. (2000). Motivation and knowledge transfer. *Organization Science, Vol. 11, No. 5* , S. 539.

Penrose, E. H. (2008). The metamorphosis of the large firm. *Organization Studies, Vol. 29, No. 8* , S. 1117-11124.

Pfeffer, J., & Nowak, P. (1976). Joint ventures and interorganizational interdependency. *Administrative Science Quarterly, Vol. 21, No. 3* , S. 398-418.

Pittaway, L., Roberston, M., Munir, K., Denyer, D., & Neely, A. (2004). *Networking and innovation in the UK: A systematic review of the literature.* United Kingdom: Advanced Institute of Management Research.

Pyka, A. (1999). *Innovation networks in economics: From the incentive based to the knowledge based approaches.* France: INRA-SERD, University Pieres Mendes.

Rallet, A., & Torre, A. (2000). *Which need for geographical proximity in innovation networks at the era of the global economy?* Paris, France: IRIS University of Paris-Dauphine.

Ranga, M. (2009). *National and regional innovation systems and policies for development: From learning regions to research intensive clusters.* Minsk, Belarus: International Conference on Knowledge-based Development.

Reagans, R., & McEvily, B. (2003). Network structure and knowledge transfer: The effects of cohesion and range. *Administrative Science Quarterly, Vol. 48, No. 2* , S. 240-267.

Reichert, S. (2006). *The rise of knowledge regions: Emerging opportunities and challanges for universities.* European University Association: EUA Publications 2006.

Rothwell, R. (1994). Towards the fifth-generation innovation process. *International marketing review, Vol. 11, No. 1* , S. 7-31.

Rycroft, R. (2007). Does cooperation absorb complexity? Innovation networks and the speed and spread of complex technological innovation. *Technological Forecasting & Social Change, Vol 74.* , S. 565-578.

Saunders, M., Lewis, P., & Thornhill, A. (2009). *Research Methods for Business Students.* Essex, UK: Pearson Educational Limited.

Simmie, J., & Kirby, M. (1998). Innovation and the theoretical base of technopole planning. *Progress in PLanning, Vol. 49.* , S. 159-198.

Spithoven, A., Clarysse, B., & Knockaert, M. (2009). *Building absorbtive capacity to organise inbound open innovation in low tech industries.* Gent, Belgium: University of Gent Working Paper, No. 606.

Sternberg, R. (2000). Innovation networks and regional development - Evidence from the european regional innovation survey. *European Planning Studies, Vol. 8, No. 4* .

Steward, F., & Conway, S. (2009). *Managing and shaping innovation.* Oxford: Oxford University Press.

Szulanski, G. (1999). *The process of knowledge transfer: A diachronic analysis of stickiness.* University of Pennsylvania: Wharton School Working Paper.

Tichy, N., Tushman, M., & Fombrun, C. (1979). Social network analysis for organisations. *Academy of Management Review, Vol. 4, No. 4* , S. 507-519.

Tödtling, F., & Trippl, M. (2005). One size fits all? Towards a differentiated regioanl innovation policy approach. *Research Policy, Vol. 34, No. 8* , S. 1203-1219.

Tran, Y., Hsuan, J., & Mahnke, V. (2011). How do innovation intermediaries add value? Insight from the new product development in fashion markets. *R&D Management* , S. 80-91.

Trochim, W. (1989). Outcome pattern matching and program theory. *Evaluation and Program Planning, Vol. 12, No. 4* , S. 355-366.

Trommsdorf, V., & Steinhoff, F. (2007). *Innovationsmarketing.* Verlag Vahlen: München.

Uotila, T. (2008). *The use of future oriented knowledge in regional innovation processes: Research on knowledge generation, transfer, and conversion.* Finland: Acta Universitatis PhD Dissertation.

Uzzi, B. (1997). Social structure and competition in interfirm networks: The paradox of embeddedness. *Administrative Science Quartely, Vol. 42* , S. 35-67.

Van de Ven, A., Polley, D., Garud, R., & Venkatraman, S. (2008). *The innovation journey.* Oxford: Oxford University Press.

Wasserman, S., & Faust, K. (1994). *Social network analysis.* New York: Cambridge University Press.

Williams, T. (2005). Cooperation by design: structure and cooperation in interorganizational networks. *Journal of Business Research, Vol 58* , S. 223-231.

Wiliamson, O. E. (1991). Comparative Economic Organization: The analysis of discrete structural alternatives. *Administrative Science Quarterly, Vol 36, No.2* , S. 269-296.

Yir, R. (2009). *Case study research : Design and methods (4th Edition).* California, USA: SAGE Publications.

Zander, U., & Kogut, B. (1996). What firms do: Coordination, identity, and learning. *Organization Science, Vol. 7, No. 5* , S. 502-518.

Zillener, A. (2011). *New Business Taxonomie. Discussion papers on economics and enrepreneurial management.* Chur, Switzerland: SIFE Schweierisches Institut für Entrepreneurship.

Appendix

A1　Innovation Actors Mentioned

The following is a comprehensive list of all actors mentioned in the external innovation activities of the firms interviewed by KARIM, listed by country of origin.

Table 25: Innovation Actor. **Source :** Own Illustration

Actor	Type	Country
World Future Energy Summit (Abu Dhabi)	Fair/exhibit	Abu Dhabi

Actor	Type	Country
NHS, National Health Service	Government Agency	England
University of Lancaster	University	England

Actor	Type	Country
Createch EU	Public Project	European Union
Enterprise Europe Network	Consultant	European Union
European Regional Development Fund	Government Agency	European Union
FP 7 - European Commission	Public Project	European Union
Project ACQUEA EU	Public Project	European Union
ZIM : Zentrales Innovationsprogramm Mittelstand	Public Project	European Union

Actor	Type	Country
Agence de l'Environnement et de la Maitrise de l'Energie (French Energy and Environment public agency)	Government Agency	France
Agence nationale de la recherche (Public funding for innovation)	Government Agency	France
Aide à la maturation de projects innovants (NGO promoting innovation)	NGO	France
Aide Pour l'Innovation (Public financial aid for innovation)	Government Agency	France
Centre Francilien de lInnovation	Government Agency	France
Centre Francilien de l'innovation (Public innovation promotion agency)	Government Agency	France
CNRS, Orsay	Government Agency	France
COFACE : Compagnie Francaise d'Assurance pour le Commerce Extérieur	Government Agency	France
Crédit d'Impot Recherche (refundable tax credit for researching firms)	Government Agency	France
CRIT : Innovation Incubator	NGO	France
Électricité de France (French public electrical company)	NGO	France
Grand Prix du Concours Lépine	Fair/exhibit	France
IEF Minerve, Orsay	Government Agency	France
IncubAlliance	Government Agency	France
Institut de recherche pour le développment	Government Agency	France
OSEO Ile-de-France	Government Agency	France
Technopole de l'Aube	NGO	France
Ubifrance	Government Agency	France

Actor	Type	Country
CyberForum e.V.	Fair/exhibit	Germany
Baden Württemberg connected (bwcon)	NGO	Germany
Bundesagentur für Arbeit (Work Agency)	Government Agency	Germany
City of Mannheim	Government Agency	Germany
Ernst & Young (Germany)	Consultant	Germany
FKM : Gesellschaft zur Freiwilligen Kontrole von Messe	NGO	Germany
Fraunhofer-Gesellschaft zur Förderung der angewandten Forschung e.V.	Research Center	Germany
Grunderstipendium : German subsidy	Government Agency	Germany
Hochschule Karlsruhe	Technical School	Germany
Leibniz-Institut für Wissensmedien (IWM)	Research Center	Germany
MBB : Mittelständische Beteiligungsgesellschaft (Venture Capital supported by Baden-Württemberg)	Government Agency	Germany
MAFINEX-Technologiezentrum	Government Agency	Germany
MBG Medien- und Filmgesellschaft Baden-Württemberg mbH	Government Agency	Germany
Online Educa	NGO	Germany
RWTH Aachen : Rheinisch-Westfälische Technische Hochschule Aachen	Technical School	Germany
Universität Leipzig	University	Germany
University Heidelberg	University	Germany
University Magdeburg	University	Germany
University Mannheim	University	Germany
University Würzburg	University	Germany

Actor	Type	Country
Dublin City University	University	Ireland
Dublin Insitute of Technology	Technical School	Ireland
Enterprise Ireland	Government Agency	Ireland
EUREKA EU	Public Project	Ireland
Queens University	University	Ireland
University College Dublin	University	Ireland

Actor	Type	Country
AO Foundation (Davos)	NGO	Switzerland
ASTRA (Bundesamt für Strassen)	Government Agency	Switzerland
École polytechnique fédérale de Lausanne EPFL	Technical School	Switzerland
Erni Consulting	Consultant	Switzerland
ETH Zürich (Swiss Federal Institute of Technology)	Technical School	Switzerland
GR Innovation : Foundation for Innovation Promotion in Graubünden	Government Agency	Switzerland
HEI ARC, Neuchatel	Technical School	Switzerland
Hochschule für Technik und Wirtschaft HTW Chur	Technical School	Switzerland
Hochschule Rapperswil	Technical School	Switzerland
NTB Buchs : Interstaatliche Hochschule für Technik Buchs	Technical School	Switzerland
Swiss Federal Commission for Innovation and Technology (CTI)	Government Agency	Switzerland
Swiss Federal Office of Energy (BFE : Bundesamt für Energie)	Government Agency	Switzerland
University of Applied Sciences Bern (Berner Fachhochschule)	University	Switzerland
Züke Consultant CH	Consultant	Switzerland

Actor	Type	Country
Veolia Environnement	Consultant	USA

A2   Data Matrix : Network Profile

The following charts detail each firm interviewed by KARIM and how their answers given correspond to the individual key network dimensions (see Chapter 5.1). These dimensions were then used to run the cluster analysis which helped to determine the existence of 4 clusters and 7 individual innovation network profiles.

Table 26: Data Matrix (Network Profile). **Source :** Own Illustration

Category	Possible Codes	1 CH	2 CH	3 CH	4 CH	5 CH	6 CH	7 CH
					Name of Firms is Confidential			
Size	0,1,2,3,4,5,6,7,8	4	6	5	5	5	6	3
Geographic Range	local, regional, national, international	national	international	international	international	international	international	national
Key Roles	custo, competitors, suppliers, consultants, univ., tech, schools, govt. body, fbro, NGO, research, other firms, data, public project, investor	customers, research institutions, other firms	customers, research institutions, governmental bodies, other firms	customers, research institutions, other firms, consultants	customers, suppliers, consultants, research institutions, other firms	governmental bodies, other firms, research centers, investors	research institutions, other firms, governmental bodies	distributors, research institutions
Actors	*list of innovation actors mentioned in interviews*	ETH, EPFL, other firms, customers	ETH, CTI, customers, other firms, FP7, BFE	University Applied Science Bern, Züke, ERNI, Frauenhofer, customers, other	HTW, suppliers, consultants, other firms, customers	CTI, FP7, GR innovation foundation, other firms, investors	FP6/7, ASTRA, Fraunhofer, Univ. of Lancaster, other firms	distributors, ETH, Profhsion school Rapperswil
Transaction Content Type	Information, goods, affect, power	information	information, affect, goods	information	information	information, affect, goods	information, affect, goods	information
Form of Innovation Supported	incremental/radical product, service, process, organization	radical product	incremental product, organization	incremental product, organization	incremental product, service, organization	organization	incremental product	incremental product
Formality	High formality, low formality	formal	formal	informal	informal	formal	formal	informal
Diversity	high diversity, low diversity	high diversity	high diversity	high diversity	high diversity	low diversity	high diversity	low diversity
Openness	open, closed	open	open	open	closed	closed	open	closed
Intensity	strong intensity, weak intensity		strong	weak		weak	strong	weak
Symmetry	symmetrical, asymmetrical	asymmetrical	asymmetrical	asymmetrical	asymmetrical	symmetrical	asymmetrical	
Type of Cooperation	knowledge generation, knowledge exploitation	knowledge generation	knowledge exploitation, knowledge generation	knowledge exploitation, knowledge generation	knowledge exploitation, knowledge generation	knowledge exploitation, knowledge generation	knowledge generation	knowledge generation
Financing	own capital, promotional money, investors	own capital	own capital, promotional money	own capital	own capital, investors	own capital, promotional money, investors	own capital, promotional money	own capital

Category	Possible Codes	8	9	10	11	12	13	14
		CH	CH	D	D	D	D	D
Size	0,1,2,3,4,5,6,7,8	5	5	2	5	5	8	8
Geographic Range	local, regional, national, international	international	international	international	regional	international	regional	international
Key Roles	custs, competitors, suppliers, consultants, unis, tech, schools, govt, bodies, fairs, NGO, research, other firms, labs, public project, investor	customers, NGOs, research institutions, testing labs, other firms	other firms, research institutions, fairs/forums, investors	other firms, distributors	research centers, fairs/forums, governmental bodies	research institutions, other firms, governmental bodies, fairs/forums, investors	customers, other firms, governmental bodies, fairs, research institutions	customers, research centers, training labs, other firms, governmental bodies
Actors	*list of innovation actors mentioned in interviews	other firms, AO foundation, Leipzig University, NTB Buchs, customers	other firms, EPFL, HE ARC, WFEI, investors	other firms, distributors	CyberForum, Baden Württemberg Connected, Mathesis, Heidelberger	MFG ONLINE EDUCA, Karlsruhe Inst., Leibniz institute for Media, investors	MFG, LESC at Karlsruhe Institute of Technology, RWTH Aachen, CyberForum, Bacton	Uni Mannheim, Uni Würzburg, other firms, Osnabrück, Uni clinic, Heidelberg, Uni
Transaction Content Type	Information, goods, affect, power	information, goods	information, goods, affect	information, goods	information, affect	information, affect, goods	information, affect	information, affect, goods
Form of Innovation Supported	incremental/radical product, service, process, organization	incremental product	radical product	incremental product, service	incremental product	incremental product	incremental product	incremental product
Formality	High formality, low formality	informal	informal	informal	informal	formal	informal	formal
Diversity	high diversity, low diversity	low diversity	high diversity	low diversity	low diversity		low diversity	high diversity
Openness	open, closed	open	open		open	open	open	open
Intensity	strong intensity, weak intensity		strong				strong	strong
Symmetry	symmetrical, asymmetrical	symmetrical	asymmetrical	symmetrical	asymmetrical		asymmetrical	asymmetrical
Type of Cooperation	knowledge generation, knowledge exploitation	knowledge generation	knowledge generation	knowledge generation	knowledge exploitation, knowledge generation	knowledge exploitation, knowledge generation	knowledge exploitation, knowledge generation	knowledge exploitation, knowledge generation
Financing	own capital, promotional money, investors	own capital	own capital, investors	own capital	own capital	own capital, investments	own capital	own capital, promotional money

Category	Possible Codes	22	23	24	25	26	27	28
		F	F	F	F	F	IRL	IRL
				Name of Firms is Confidential				
Size	0,1,2,3,4,5,6,7,8	4	6	5	8	5	5	8
Geographic Range	local, regional, national, international	international	national	national	national	national	international	international
Key Roles	custs, competitors, suppliers, consultants, univ., tech. schools, govt. body, firm, NGO, research, other firms, labs, public project, investor	governmental bodies, other firms, consultants	customers, distributors, government bodies, consultants	governmental bodies, research institution, other firms	research institutions, governmental bodies, fairs/forums, other	governmental bodies, research institutions, other firms, investors	research institutions, governmental bodies, customers	governmental bodies, research centre, customers
Actors	*list of innovation actors mentioned in interviews	EU Project Cluster ACOLSEAL OSEO (France), other firms, EEN Enterprise	CIR, API, COFACE, UBI France, customer, distributors	ANR, AIMA, CNRS, IEF, CFI	IRD, OSEO, CRIT, INCUBALLIANCE, other firms, technopole, grand prix, CEI	CFI, OSEO, ADAME, EDF, investors	Enterprise Ireland, CREST, customers, FP7, other firms	Enterprise Ireland, EUREKA, FP7, QUESTOR (Queen's University Belfast)
Transaction Content Type	information, goods, effect, power	information, a ffect, goods	information, goods	information, goods	information, goods	information, goods	information, a ffect, goods	information, a ffect, goods
Form of Innovation Supported	incremental/radical product, service, process, organization	incremental product, organization	incremental product, organization	incremental product	radical product	incremental product	incremental product	incremental product, service
Formality	High formality, low formality	formal		formal	formal	formal	formal	formal
Diversity	high diversity, low diversity	high diversity	low diversity	high diversity	high diversity	high diversity	high diversity	high diversity
Openness	open, closed	open		open	open	open	open	open
Intensity	strong intensity, weak intensity			strong	strong	strong	strong	strong
Symmetry	symmetrical, asymmetrical	asymmetrical		asymmetrical	asymmetrical	asymmetrical	asymmetrical	asymmetrical
Type of Cooperation	knowledge generation, knowledge exploitation	knowledge exploitation, knowledge generation	knowledge exploitation, knowledge generation	knowledge exploitation, knowledge generation	knowledge exploitation	knowledge exploitation, knowledge generation	knowledge generation	knowledge generation
Financing	own capital, promotional money, investors	promotional money	promotional money	own capital, promotional money	promotional money	own capital, promotional money, investors	own capital, promotional money	own capital, promotional money

Category	Possible Codes	15 D	16 D	17 ENG	18 ENG	19 ENG	20 ENG	21 ENG
					Name of Firms is Confidential			
Size	0,1,2,3,4,5,6,7,8	8	3	0	2	4	2	1
Geographic Range	local, regional, national, international	international	national	local	national	international	national	regional
Key Roles	custs, competitors, suppliers, consultants, univ, tech, schools, govt body, fbm, NGO, research, other firms, labs, public project, investor	customers, research centers, consultant, other firms, governmental bodies	customers, research centers, NGOs, governmental bodies		suppliers, other firms	other firm, governmental bodies, NGOs, research institutions	research institution, governmental bodies	research institution
Actors	*list of innovation actors mentioned in interviews	Uni Heidelberg, FKM, Mannheim city, Createch, other firms, investors	German Founders Stipendium, customers, Karlsruhe Institute of Tech, EU projects?		suppliers, other firms	FP7, ERDF, other firms, research institutions	InfoLab21 Lancaster Univ, NHS (National Health Service)	Lancaster University
Transaction Content Type	information, goods, affect, power	information, affect, goods	information, affect		information	information, affect	information, affect	information
Form of Innovation Supported	incremental/radical product, service, process, organisation	incremental product	radical product		service	service	service	incremental product
Formality	High formality, low formality	informal	informal			formal	informal	informal
Diversity	high diversity, low diversity	low diversity	low diversity			high diversity	low diversity	low diversity
Openness	open, closed	open	open	closed	closed	open		open
Intensity	strong intensity, weak intensity	strong			weak	strong		weak
Symmetry	symmetrical, asymmetrical	asymmetrical	symmetrical		symmetrical	asymmetrical		asymmetrical
Type of Cooperation	knowledge generation, knowledge exploitation	knowledge exploitation, knowledge generation	knowledge exploitation, knowledge generation		knowledge generation	knowledge generation	knowledge generation	knowledge generation
Financing	own capital, promotional money, investors	own capital, promotional money, investors	own capital, promotional money		own capital	own capital	own capital	own capital

A3 Innovation Networks

The following are graphical representations of the specific and exact innovation networks found from the KARIM Data Set, used to help create the network profiles found in the main paper.

KEY

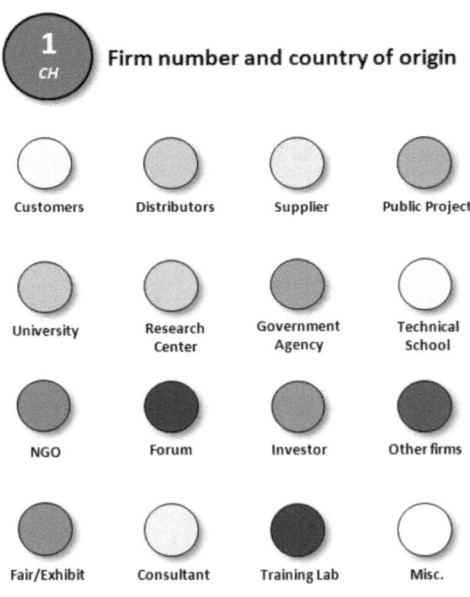

Customers	Distributors	Supplier	Public Project
University	Research Center	Government Agency	Technical School
NGO	Forum	Investor	Other firms
Fair/Exhibit	Consultant	Training Lab	Misc.

Firm number and country of origin

A3.1 Innovation Network: Knowledge and Learning

Figure 10: Knowledge and Learning Networks. **Source :** Own Illustration

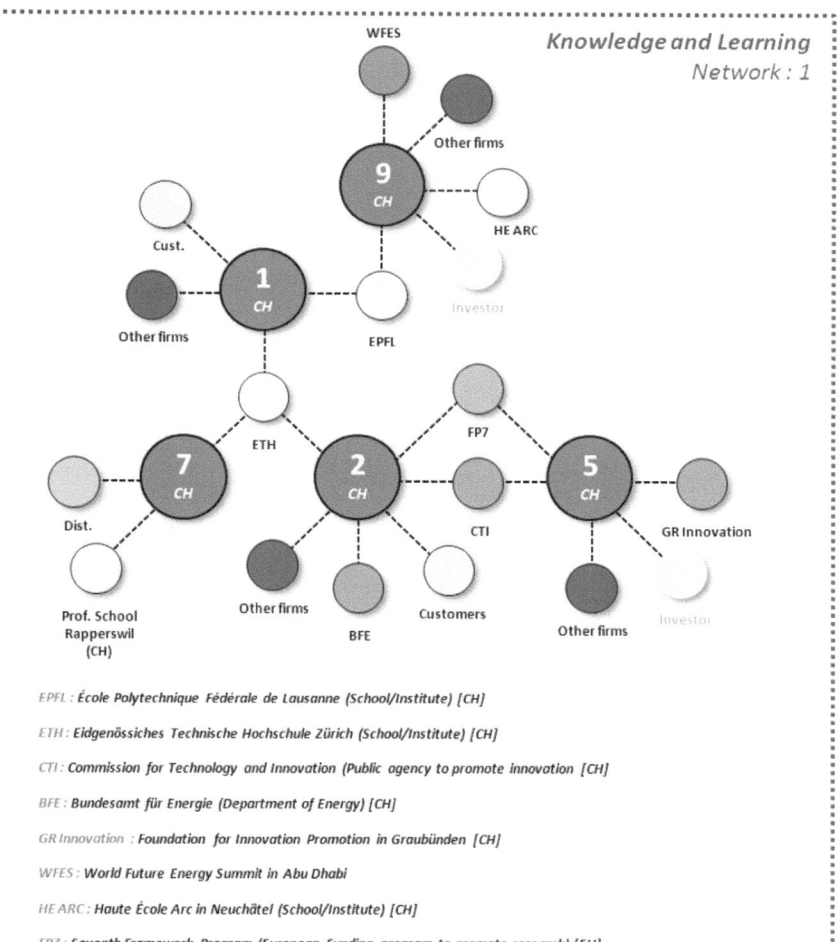

EPFL : *École Polytechnique Fédérale de Lausanne (School/Institute) [CH]*

ETH : *Eidgenössiches Technische Hochschule Zürich (School/Institute) [CH]*

CTI : *Commission for Technology and Innovation (Public agency to promote innovation [CH]*

BFE : *Bundesamt für Energie (Department of Energy) [CH]*

GR Innovation : *Foundation for Innovation Promotion in Graubünden [CH]*

WFES : *World Future Energy Summit in Abu Dhabi*

HE ARC : *Haute École Arc in Neuchâtel (School/Institute) [CH]*

FP7 : *Seventh Framework Program (European Funding program to promote research) [EU]*

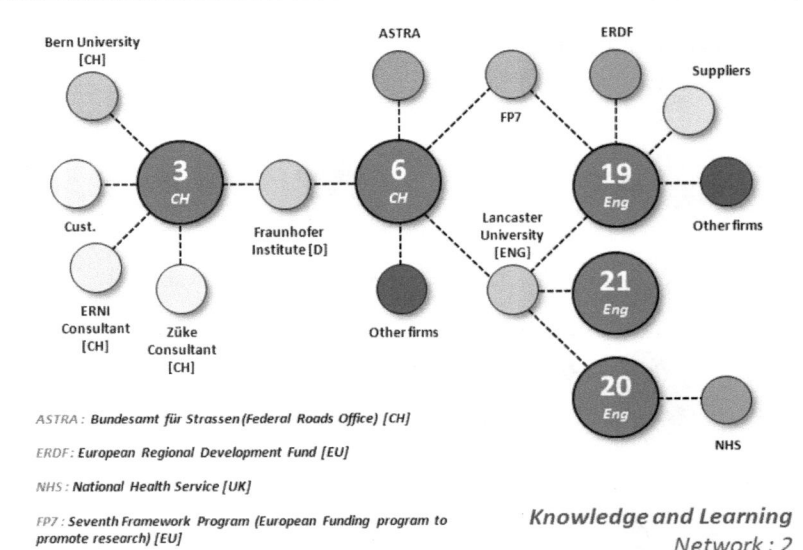

ASTRA : Bundesamt für Strassen (Federal Roads Office) [CH]

ERDF: European Regional Development Fund [EU]

NHS : National Health Service [UK]

FP7 : Seventh Framework Program (European Funding program to promote research) [EU]

Knowledge and Learning
Network : 2

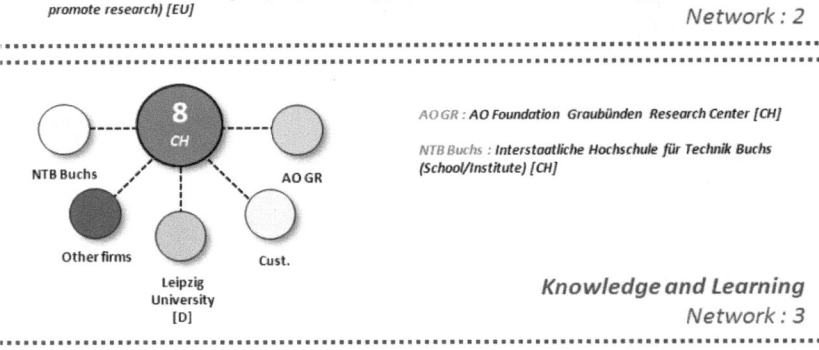

AO GR : AO Foundation Graubünden Research Center [CH]

NTB Buchs : Interstaatliche Hochschule für Technik Buchs (School/Institute) [CH]

Knowledge and Learning
Network : 3

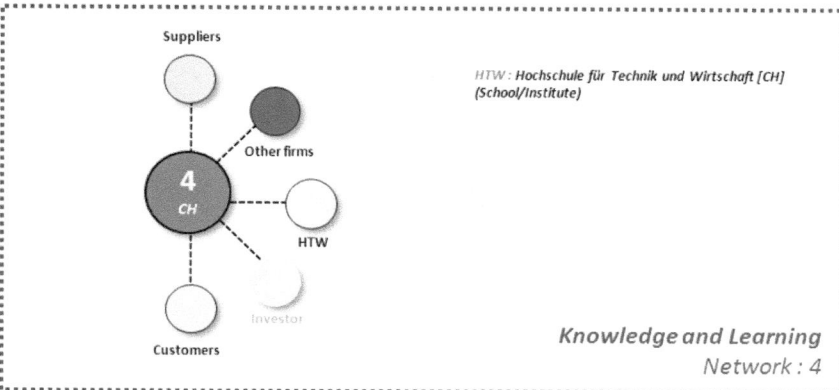

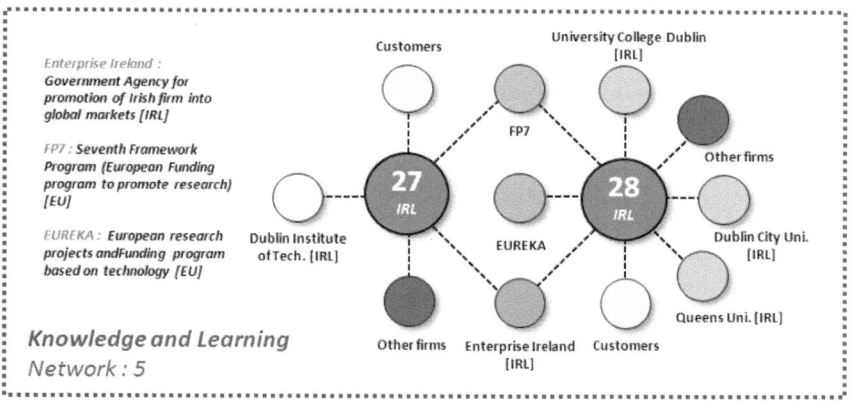

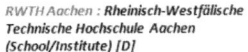

RWTH Aachen : *Rheinisch-Westfälische Technische Hochschule Aachen (School/Institute) [D]*

MFG : *Medien –und filmgesellschaft (Governement Agency promoting media technology) [D]*

Cyberforum : *High-tech firm network [D]*

KIT : *Karlsruhe Institute für Technologie (School/Institute) [D]*

bwcon : *Baden-Württemberg Connected (Technology network of firms) [D]*

Online Educa : *E-learning conference [D]*

MAFINEX : *Technology center [D]*

GS: Gründerstipendium : *German subsidy for founding a firm [D]*

BFE : *Bundesagentur für Arbeit (German work agency) [D]*

MBG : *Mittelständische Beteiligungsgesellschaft (Public VC) [D]*

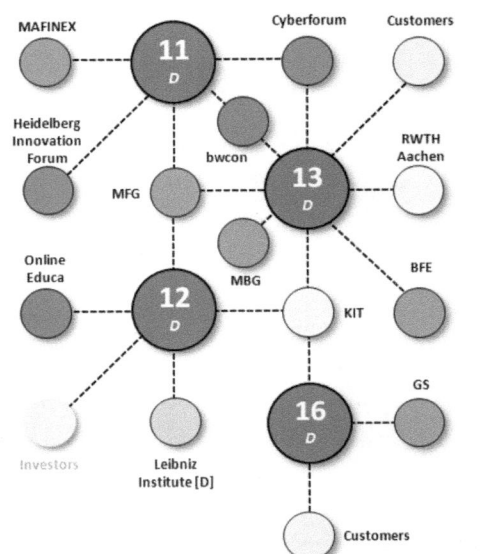

Knowledge and Learning
Network : 6

ZIM : *Zentrales Innovationsprogramm Mittelstand (German reseearch projects) [D]*

Createch : *European Research Project [EU]*

FKM : *Gesellschaft zur Freiwilligen Kontrole von Messe (Society of Fair Monitoring and controlling) [D]*

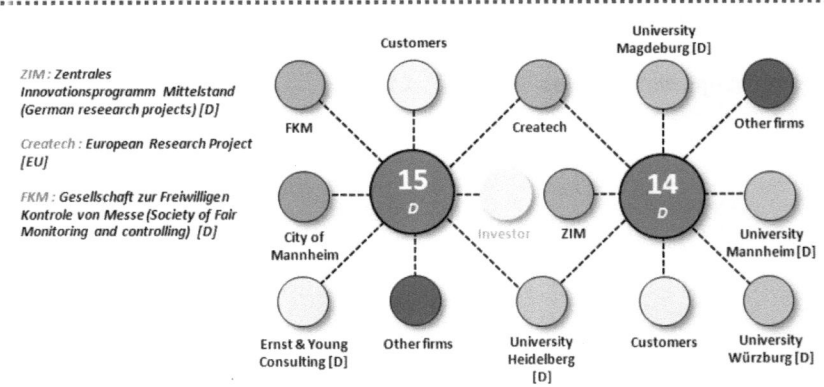

Knowledge and Learning
Network : 7

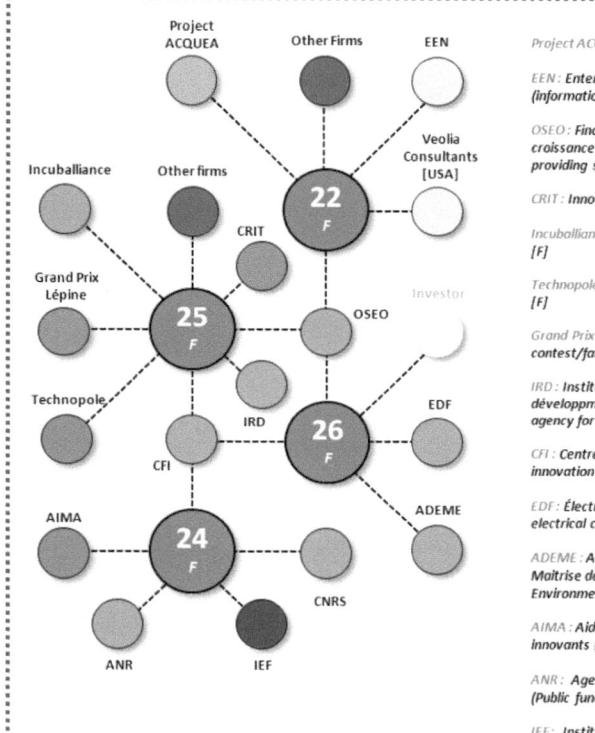

Project ACQUEA : *EU partner project [EU]*

EEN : Enterprise Europe Network (information and consultancy network) [EU]

OSEO : Financement de l'innovation et de la croissance des PME (Public organization providing strategic and financial support) [F]

CRIT : Innovation Incubator [F]

Incuballiance : Public Innovation Incubator [F]

Technopole de l'Aube : Innovation technopole [F]

Grand Prix du Concours Lépine : Innovation contest/fair [F]

IRD : Institut de recherche pour le développment (Public innovation promotion agency for Southern regions) [F]

CFI : Centre Francillien de l'innovation (Public innovation promotion agency) [F]

EDF : Électricité de France (French public electrical company) [F]

ADEME : Agence de l'Environnement et de la Maitrise de l'Energie (French Energy and Environment public agency) [F]

AIMA : Aide à la maturation de projects innovants (NGO promoting innovation) [F]

ANR : Agence nationale de la recherche (Public funding for innovation) [F]

IEF : Institut d'Electronique Fondamentale (Private research lab) [F]

CNRS : Centre national de la recherche scientifique (French national research center) [F]

Knowledge and Learning
Network : 8

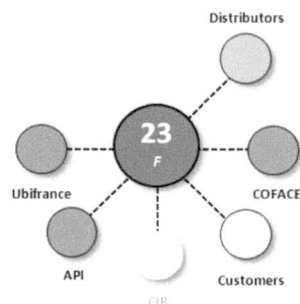

COFACE : Compagnie Francaise d'Assurance pour le Commerce Extérieur [F]

Ubifrance : French Agency for exporting and consulting aid [F]

CIR : Crédit d'Impot Recherche (refundable tax credit for researching firms) [F]

API : Aide Pour l'Innovation (Public financial aid for innovation) [F]

Knowledge and Learning
Network : 9

A3.2 Innovation Networks: Financial Procurement

Figure 11: Financial Procurement Networks. **Source :** Own Illustration

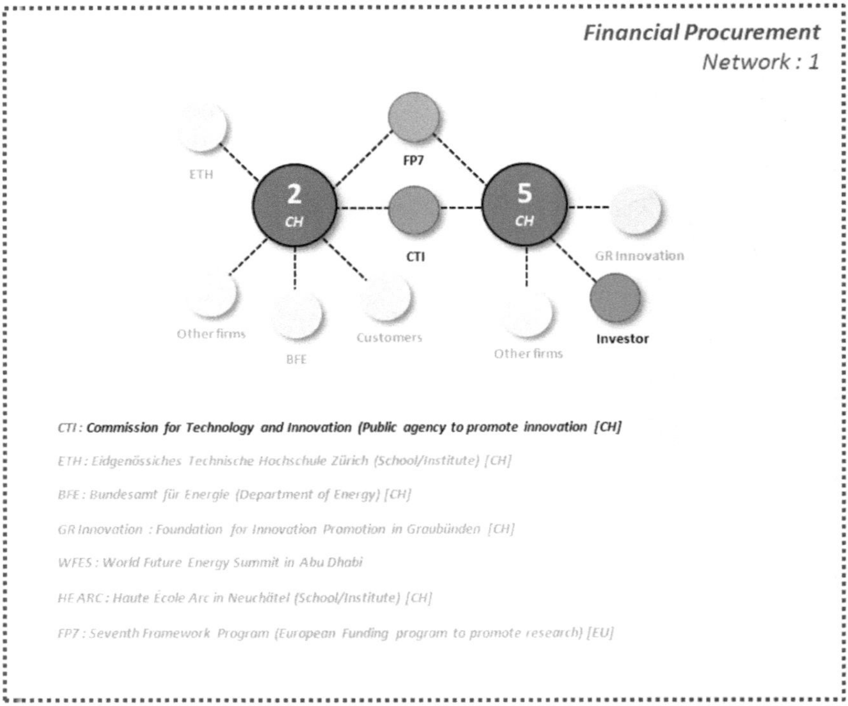

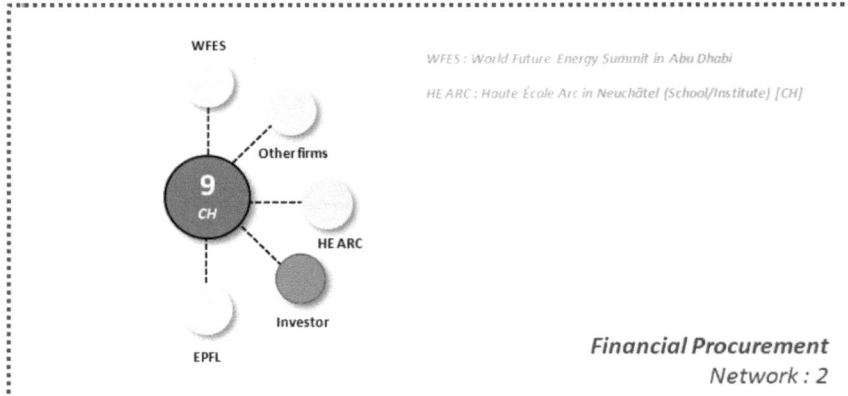

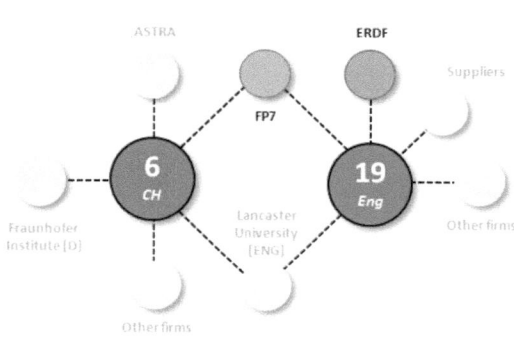

ERDF : European Regional Development Fund [EU]

FP7 : Seventh Framework Program (European Funding program to promote research) [EU]

ASTRA : Bundesamt für Strassen (Federal Roads Office) [CH]

NHS : National Health Service [UK]

Financial Procurement
Network : 3

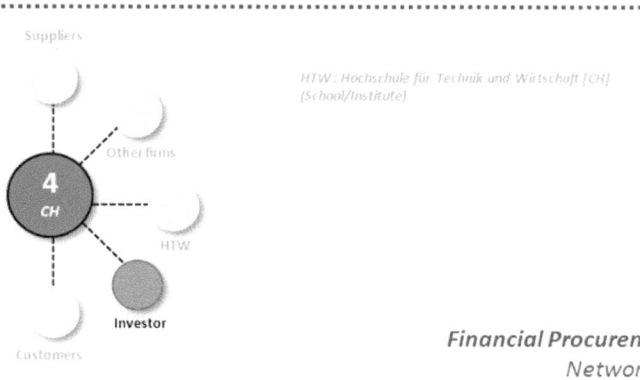

Financial Procurement
Network : 4

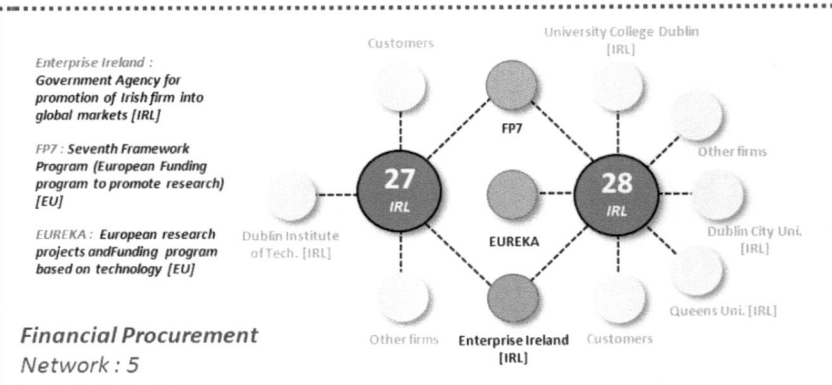

**Financial Procurement
Network : 5**

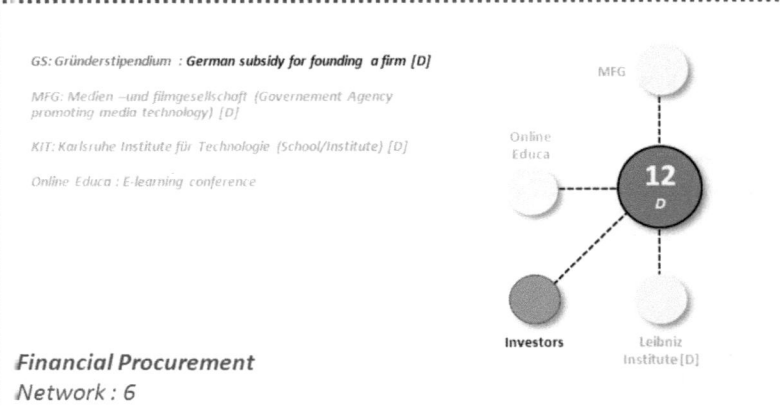

**Financial Procurement
Network : 6**

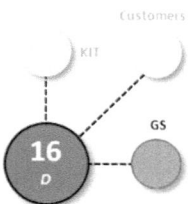

GS: *Gründerstipendium* : **German subsidy for founding a firm [D]**

KIT: *Karlsruhe Institute für Technologie (School/Institute) [D]*

Financial Procurement
Network : 7

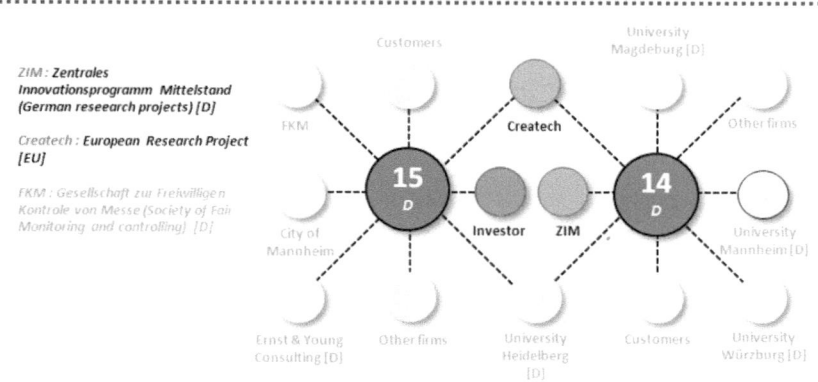

ZIM : **Zentrales Innovationsprogramm Mittelstand (German reseearch projects) [D]**

Createch : **European Research Project [EU]**

FKM : *Gesellschaft zur Freivilligen Kontrole von Messe (Society of Fair Monitoring and controlling) [D]*

Financial Procurement
Network : 8

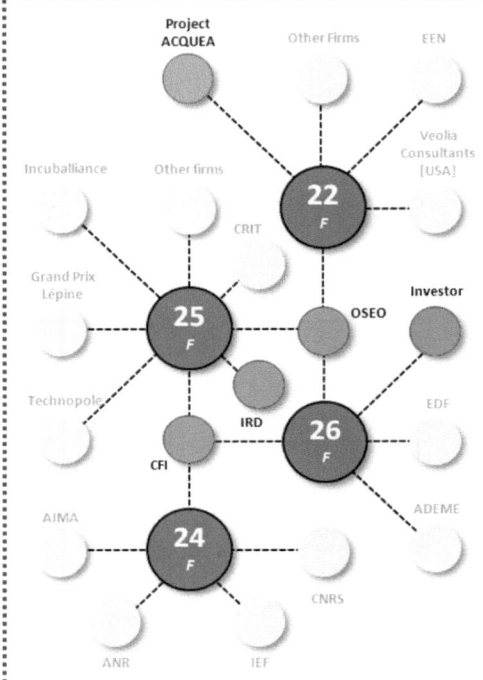

Financial Procurement
Network : 9

Project ACQUEA : **EU partner project [EU]**

OSEO : **Financement de l'innovation et de la croissance des PME (Public organization providing strategic and financial support) [F]**

IRD : **Institut de recherche pour le développment (Public innovation promotion agency for Southern regions) [F]**

CFI : **Centre Francillien de l'innovation (Public innovation promotion agency) [F]**

CRIT : Innovation Incubator [F]

Incuballiance : Public Innovation Incubator [F]

Technopole de l'Aube: Innovation technopole [F]

Grand Prix du Concours Lépine : Innovation contest/fair [F]

EEN : Enterprise Europe Network (information and consultancy network) [EU]

EDF : Électricité de France (French public electrical company) [F]

ADEME : Agence de l'Environnement et de la Maîtrise de l'Energie (French Energy and Environment public agency) [F]

AIMA : Aide à la maturation de projects innovants (NGO promoting innovation) [F]

ANR : Agence nationale de la recherche (Public funding for innovation) [F]

IEF : Institut d'Electronique Fondamentale (Private research lab) [F]

CNRS : Centre national de la recherche scientifique (French national research center) [F]

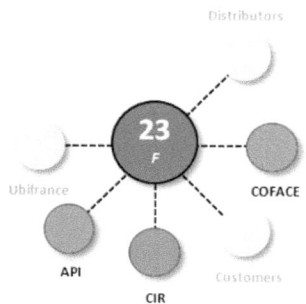

COFACE : **Compagnie Francaise d'Assurance pour le Commerce Extérieur [F]**

CIR : **Crédit d'Impot Recherche (refundable tax credit for researching firms) [F]**

API : **Aide Pour l'Innovation (Public financial aid for innovation) [F]**

Ubifrance : French Agency for exporting and consulting aid [F]

Financial Procurement
Network : 10

BFE : **Bundesagentur für Arbeit (German work agency) [D]**

MBG: **Mittelständische Beteiligungsgesellschaft (Public VC) [D]**

RWTH Aachen : Rheinisch-Westfälische Technische Hochschule Aachen (School/Institute) [D]

MFG : Medien –und filmgesellschaft (Governement Agency promoting media technology) [D]

Cyberforum : High-tech firm network [D]

KIT : Karlsruhe Institute für Technologie (School/Institute) [D]

bwcon : Baden-Württemberg Connected (Technology network of firms) [D]

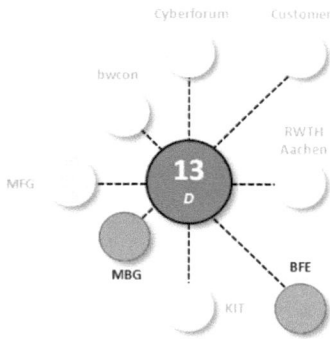

Financial Procurement
Network : 11

A3 3   Innovation Networks: Public-Private Cooperation

Figure 12: Public-Private Cooperation Networks. **Source** : Own Illustration

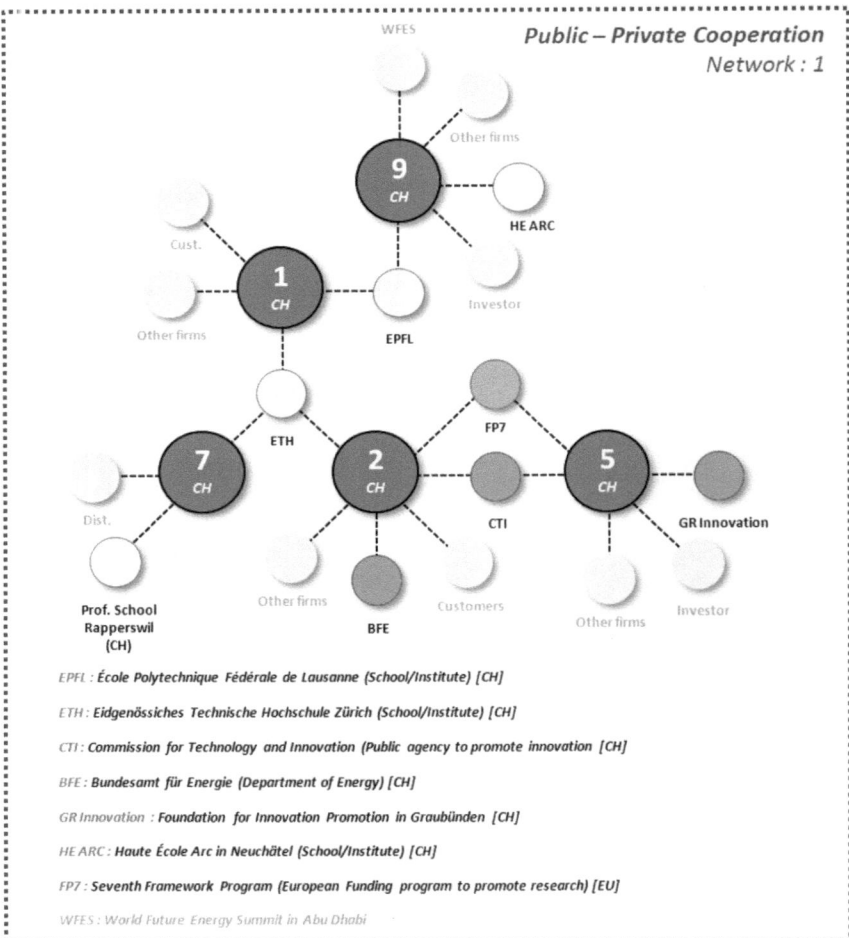

Public – Private Cooperation
Network : 1

EPFL : École Polytechnique Fédérale de Lausanne (School/Institute) [CH]

ETH : Eidgenössiches Technische Hochschule Zürich (School/Institute) [CH]

CTI : Commission for Technology and Innovation (Public agency to promote innovation [CH]

BFE : Bundesamt für Energie (Department of Energy) [CH]

GR Innovation : Foundation for Innovation Promotion in Graubünden [CH]

HE ARC : Haute École Arc in Neuchâtel (School/Institute) [CH]

FP7 : Seventh Framework Program (European Funding program to promote research) [EU]

WFES : World Future Energy Summit in Abu Dhabi

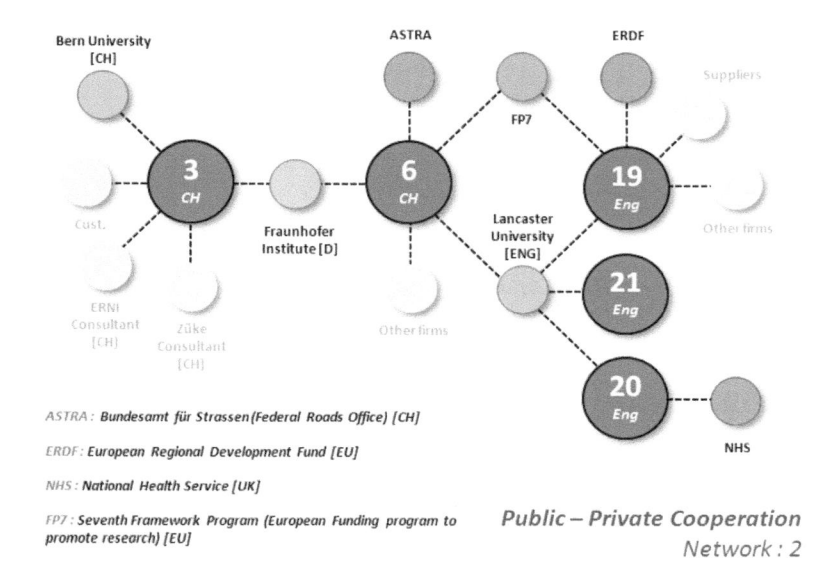

Bern University [CH]

ASTRA

ERDF

Suppliers

3 CH

FP7

6 CH

19 Eng

Cust.

Fraunhofer Institute [D]

Lancaster University [ENG]

Other firms

ERNI Consultant [CH]

Züke Consultant [CH]

Other firms

21 Eng

20 Eng

NHS

ASTRA : Bundesamt für Strassen (Federal Roads Office) [CH]

ERDF : European Regional Development Fund [EU]

NHS : National Health Service [UK]

FP7 : Seventh Framework Program (European Funding program to promote research) [EU]

Public – Private Cooperation
Network : 2

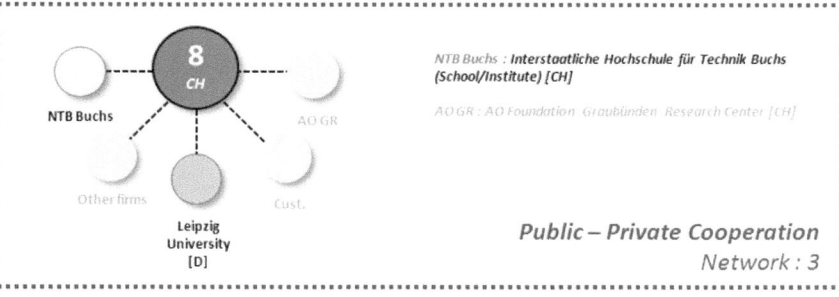

8 CH

NTB Buchs : Interstaatliche Hochschule für Technik Buchs (School/Institute) [CH]

NTB Buchs

AO GR

AO GR : AO Foundation Graubünden Research Center [CH]

Other firms

Cust.

Leipzig University [D]

Public – Private Cooperation
Network : 3

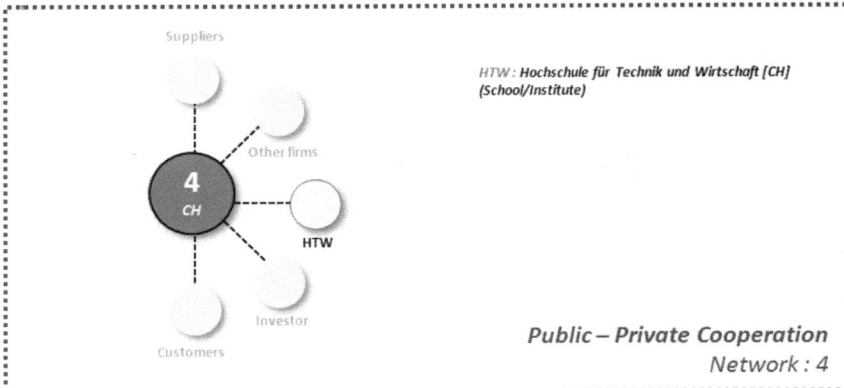

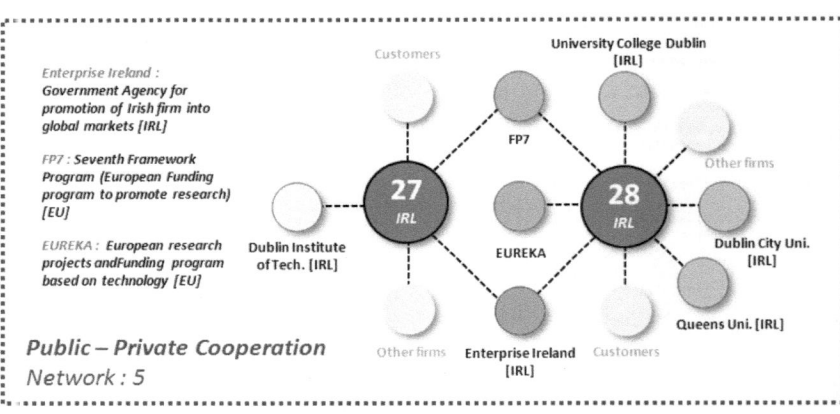

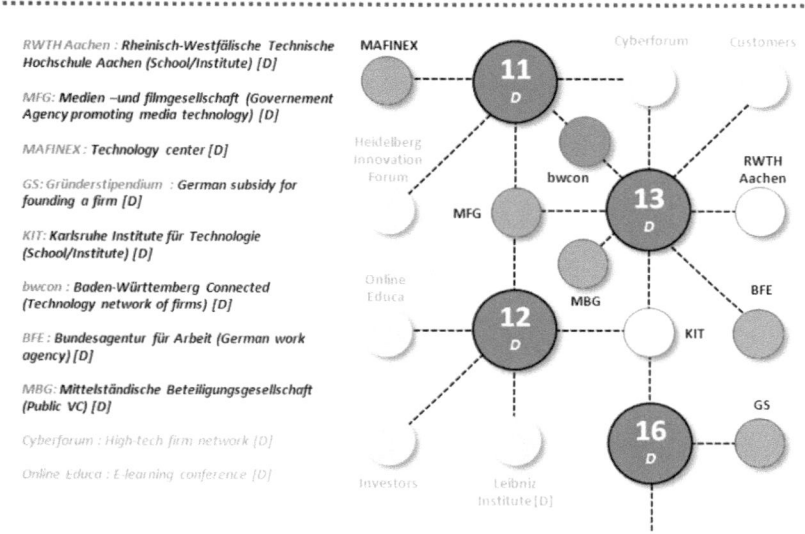

RWTH Aachen : **Rheinisch-Westfälische Technische Hochschule Aachen (School/Institute) [D]**

MFG: **Medien –und filmgesellschaft (Governement Agency promoting media technology) [D]**

MAFINEX : **Technology center [D]**

GS: Gründerstipendium : **German subsidy for founding a firm [D]**

KIT: **Karlsruhe Institute für Technologie (School/Institute) [D]**

bwcon : **Baden-Württemberg Connected (Technology network of firms) [D]**

BFE : **Bundesagentur für Arbeit (German work agency) [D]**

MBG: **Mittelständische Beteiligungsgesellschaft (Public VC) [D]**

Cyberforum : High-tech firm network [D]

Online Educa : E-learning conference [D]

Public – Private Cooperation
Network : 6

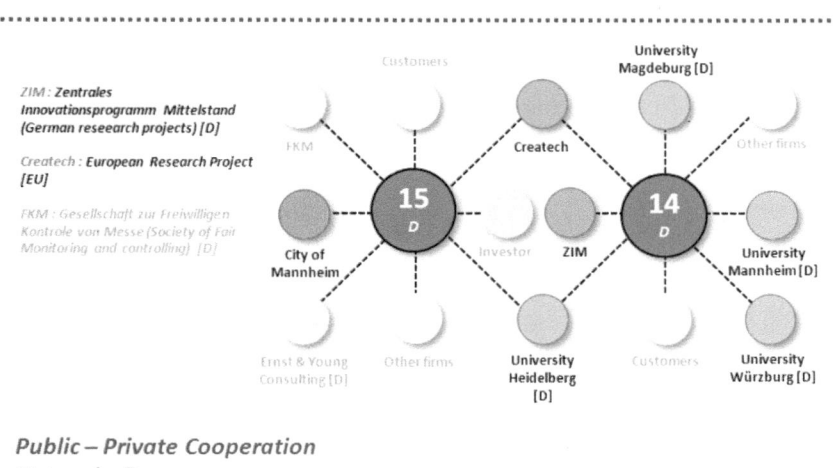

ZIM : **Zentrales Innovationsprogramm Mittelstand (German reseearch projects) [D]**

Createch : **European Research Project [EU]**

FKM : Gesellschaft zur Freiwilligen Kontrole von Messe (Society of Fair Monitoring and controlling) [D]

Public – Private Cooperation
Network : 7

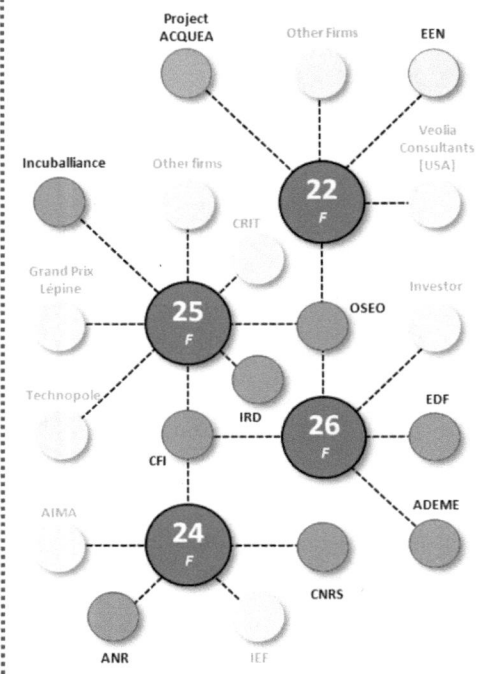

Project ACQUEA : EU partner project [EU]

EEN : Enterprise Europe Network (information and consultancy network) [EU]

OSEO : Financement de l'innovation et de la croissance des PME (Public organization providing strategic and financial support) [F]

IRD : Institut de recherche pour le développment (Public innovation promotion agency for Southern regions) [F]

CFI : Centre Francillien de l'innovation (Public innovation promotion agency) [F]

EDF : Électricité de France (French public electrical company) [F]

ADEME : Agence de l'Environnement et de la Maîtrise de l'Energie (French Energy and Environment public agency) [F]

ANR : Agence nationale de la recherche (Public funding for innovation) [F]

CNRS : Centre national de la recherche scientifique (French national research center) [F]

Incuballiance : Public Innovation Incubator [F]

AIMA : Aide à la maturation de projects innovants (NGO promoting innovation) [F]

CRIT : Innovation Incubator [F]

IEF : Institut d'Electronique Fondamentale (Private research lab) [F]

Technopole de l'Aube: Innovation technopole [F]

Grand Prix du Concours Lépine : Innovation contest/fair [F]

Public – Private Cooperation
Network : 8

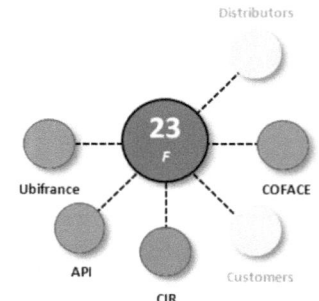

COFACE : Compagnie Francaise d'Assurance pour le Commerce Extérieur [F]

Ubifrance : French Agency for exporting and consulting aid [F]

CIR : Crédit d'Impot Recherche (refundable tax credit for researching firms) [F]

API : Aide Pour l'Innovation (Public financial aid for innovation) [F]

Knowledge and Learning
Network : 9

A3.4 Innovation Networks: Vertical Integration

Figure 13: Vertical Integrations Networks. **Source :** Own Illustration

EPFL : École Polytechnique Fédérale de Lausanne
(School/Institute) [CH]

ETH : Eidgenössiches Technische Hochschule Zürich
(School/Institute) [CH]

CTI : Commission for Technology and Innovation
(Public agency to promote innovation [CH]

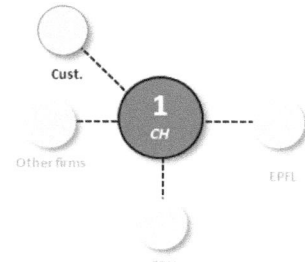

Vertical Integration
Network : 1

ETH : Eidgenössiches Technische Hochschule Zürich
(School/Institute) [CH]

CTI : Commission for Technology and Innovation
(Public agency to promote innovation [CH]

BFE : Bundesamt für Energie (Department of Energy)
[CH]

FP7 : Seventh Framework Program (European Funding
program to promote research) [EU]

Vertical Integration
Network : 2

ETH : Eidgenössiches Technische Hochschule Zürich
(School/Institute) [CH]

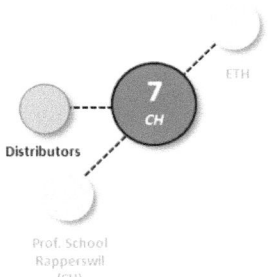

Vertical Integration
Network : 3

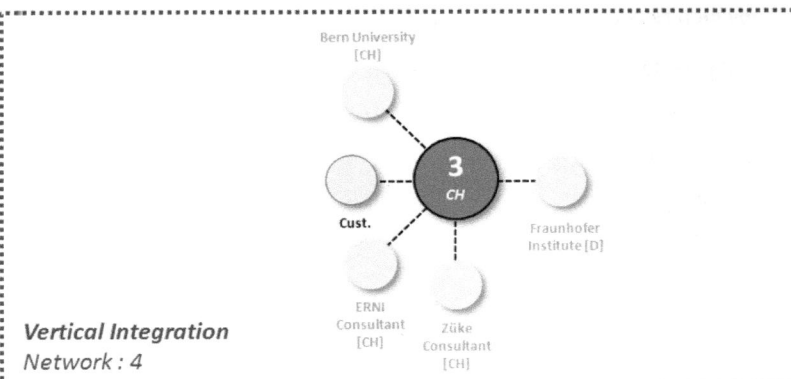

Bern University
[CH]

3
CH

Cust.

Fraunhofer
Institute [D]

ERNI
Consultant
[CH]

Züke
Consultant
[CH]

Vertical Integration
Network : 4

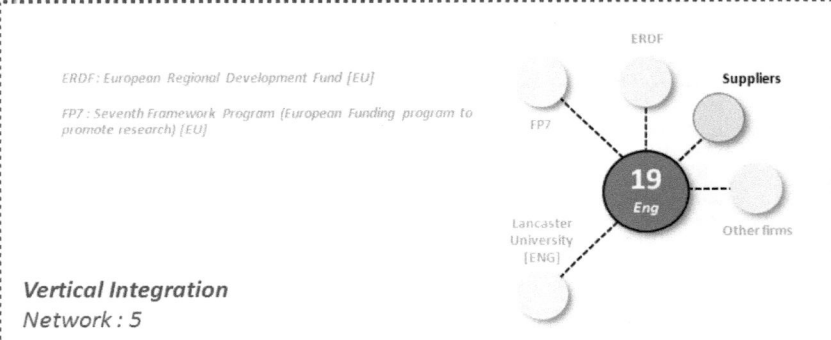

ERDF

ERDF : European Regional Development Fund [EU]

*FP7 : Seventh Framework Program (European Funding program to
promote research) [EU]*

Suppliers

FP7

19
Eng

Lancaster
University
[ENG]

Other firms

Vertical Integration
Network : 5

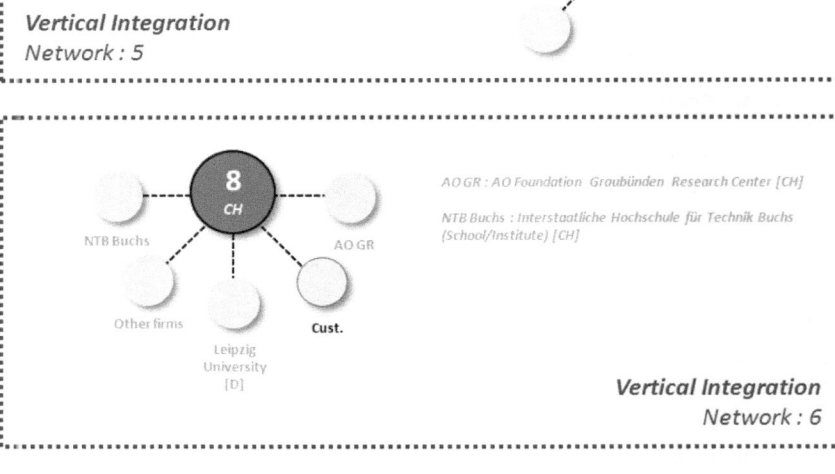

8
CH

NTB Buchs

AO GR

Other firms

Cust.

Leipzig
University
[D]

AO GR : AO Foundation Graubünden Research Center [CH]

*NTB Buchs : Interstaatliche Hochschule für Technik Buchs
(School/Institute) [CH]*

Vertical Integration
Network : 6

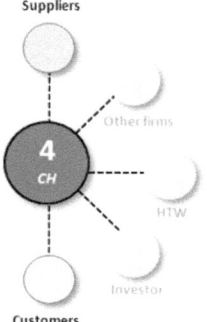

Suppliers

Other firms

4
CH

HTW

Investor

Customers

HTW : Hochschule für Technik und Wirtschaft [CH] (School/Institute)

Vertical Integration
Network : 7

Customers

FP7

27
IRL

Dublin Institute of Tech. [IRL]

Other firms   Enterprise Ireland [IRL]

Enterprise Ireland : Government Agency for promotion of Irish firm into global markets [IRL]

FP7 : Seventh Framework Program (European Funding program to promote research) [EU]

Vertical Integration
Network : 8

Enterprise Ireland : Government Agency for promotion of Irish firm into global markets [IRL]

FP7 : Seventh Framework Program (European Funding program to promote research) [EU]

EUREKA : European research projects andFunding program based on technology [EU]

University College Dublin [IRL]

FP7

Other firms

28
IRL

Dublin City Uni. [IRL]

EUREKA

Queens Uni. [IRL]

Enterprise Ireland [IRL]   Customers

Vertical Integration
Network : 9

RWTH Aachen : Rheinisch-Westfälische Technische Hochschule
Aachen (School/Institute) [D]

MFG: Medien –und filmgesellschaft (Governement Agency
promoting media technology) [D]

Cyberforum : High-tech firm network [D]

KIT: Karlsruhe Institute für Technologie (School/Institute) [D]

bwcon : Baden-Württemberg Connected (Technology network
of firms) [D]

BFE : Bundesagentur für Arbeit (German work agency) [D]

MBG: Mittelständische Beteiligungsgesellschaft (Public VC) [D]

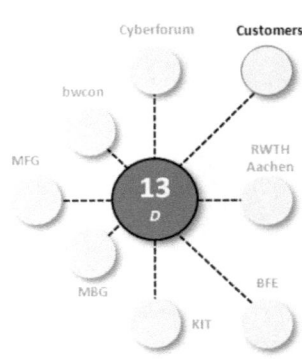

Vertical Integration
Network : 10

KIT: Karlsruhe Institute für Technologie
(School/Institute) [D]

GS: Gründerstipendium : German subsidy for
founding a firm [D]

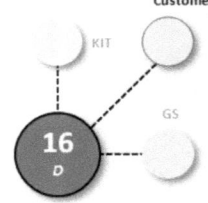

Vertical Integration
Network : 11

Vertical Integration
Network : 12

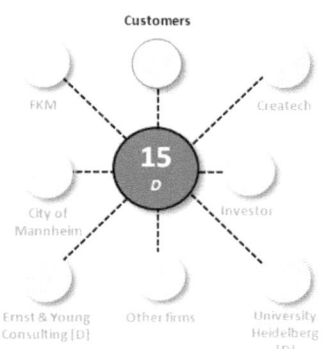

Customers

FKM

Createch

15
D

City of Mannheim

Investor

Ernst & Young Consulting [D]

Other firms

University Heidelberg [D]

Createch : European Research Project [EU]

FKM : Gesellschaft zur Freiwilligen Kontrole von Messe (Society of Fair Monitoring and controlling) [D]

Vertical Integration
Network : 13

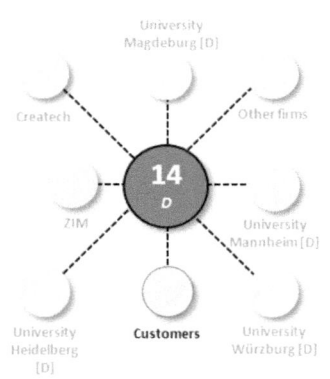

University Magdeburg [D]

Createch

Other firms

14
D

ZIM

University Mannheim [D]

University Heidelberg [D]

Customers

University Würzburg [D]

ZIM : Zentrales Innovationsprogramm Mittelstand (German research projects) [D]

Createch : European Research Project [EU]

Vertical Integration
Network : 14

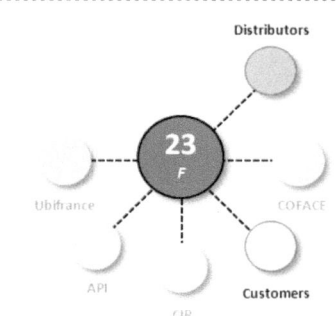

Distributors

23
F

Ubifrance

COFACE

API

Customers

CIR

COFACE : Compagnie Francaise d'Assurance pour le Commerce Extérieur [F]

Ubifrance : French Agency for exporting and consulting aid [F]

CIR : Crédit d'Impot Recherche (refundable tax credit for researching firms) [F]

API : Aide Pour l'Innovation (Public financial aid for innovation) [F]

Vertical Integration
Network : 15

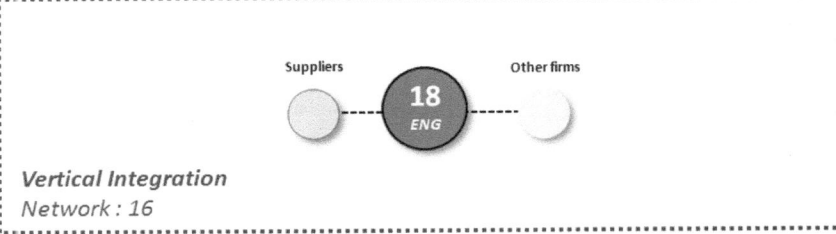

Suppliers
18
ENG
Other firms
Vertical Integration
Network : 16

A3.5 Innovation Networks: Regional Clusters

Figure 14: Regional Clusters Networks. **Source :** Own Illustration

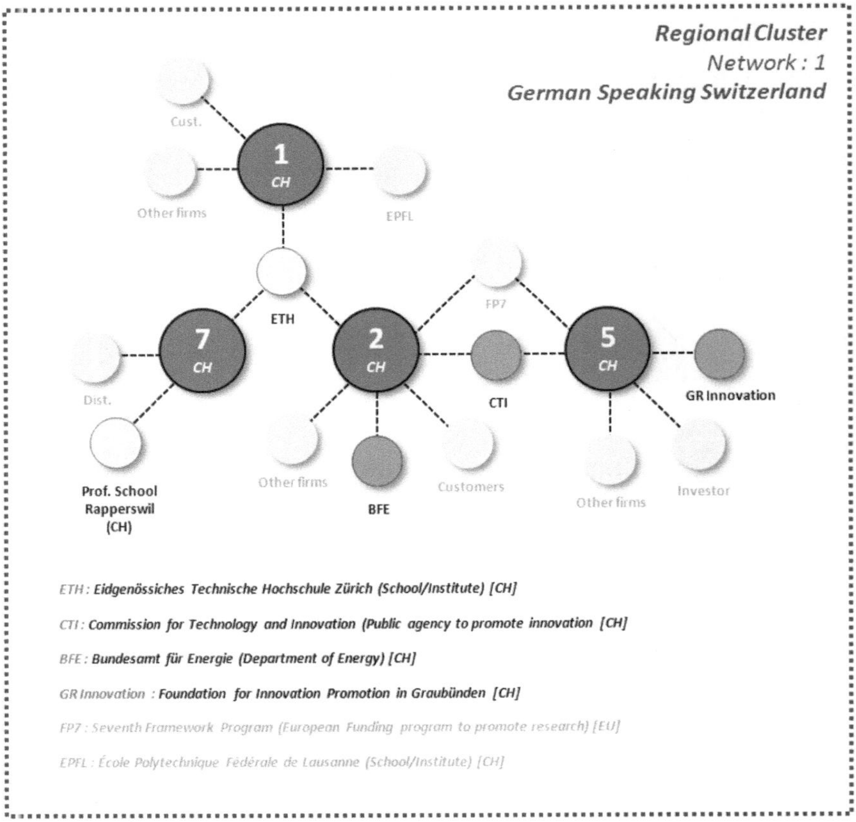

Regional Cluster
Network : 1
German Speaking Switzerland
Cust.
1
CH
Other firms
EPFL
ETH
FP7
7
CH
2
CH
CTI
5
CH
GR Innovation
Dist.
Prof. School Rapperswil (CH)
Other firms
BFE
Customers
Other firms
Investor
ETH : Eidgenössiches Technische Hochschule Zürich (School/Institute) [CH]
CTI : Commission for Technology and Innovation (Public agency to promote innovation [CH]
BFE : Bundesamt für Energie (Department of Energy) [CH]
GR Innovation : Foundation for Innovation Promotion in Graubünden [CH]
FP7 : Seventh Framework Program (European Funding program to promote research) [EU]
EPFL : École Polytechnique Fédérale de Lausanne (School/Institute) [CH]

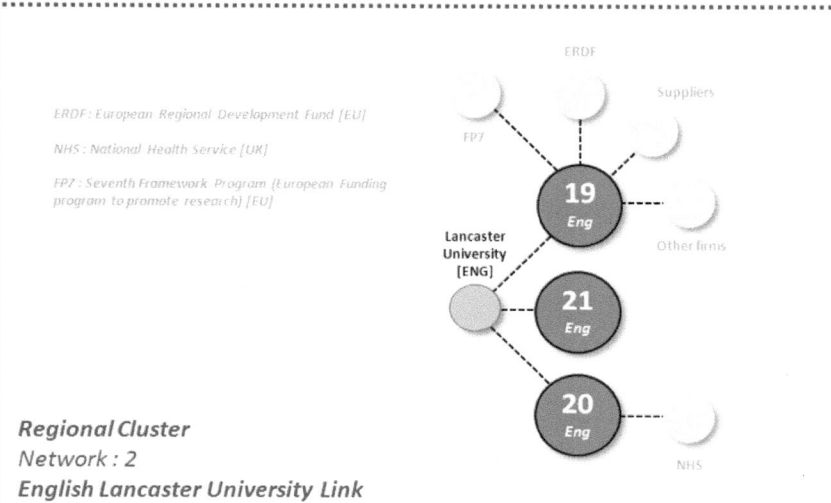

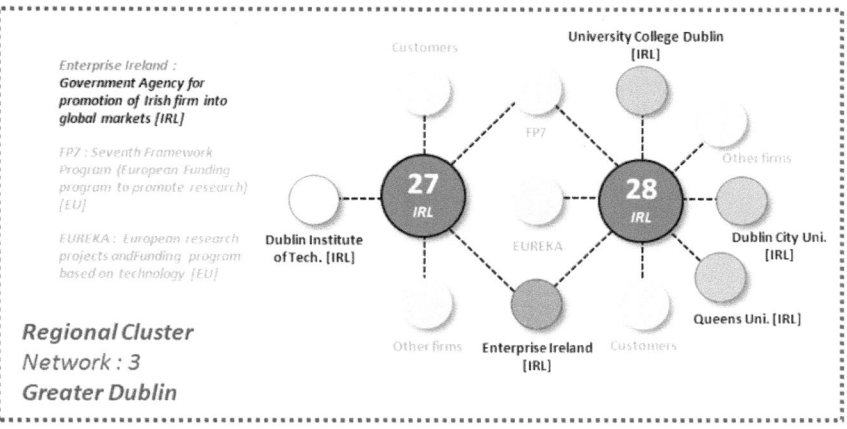

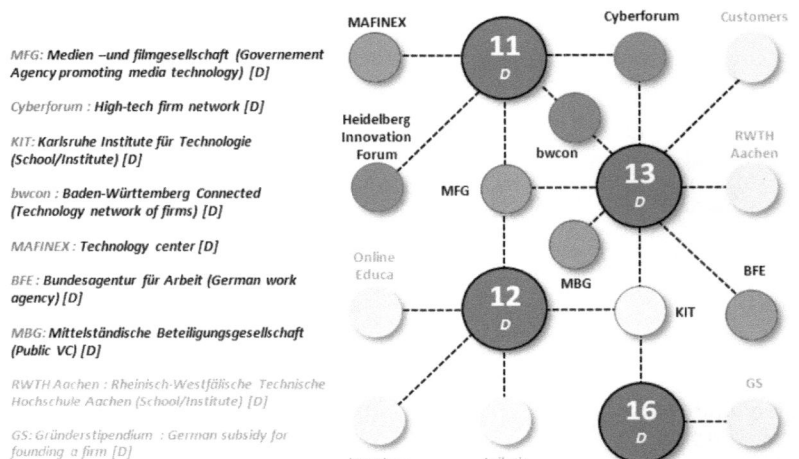

MFG: Medien –und filmgesellschaft (Governement Agency promoting media technology) [D]

Cyberforum : High-tech firm network [D]

KIT: Karlsruhe Institute für Technologie (School/Institute) [D]

bwcon : Baden-Württemberg Connected (Technology network of firms) [D]

MAFINEX : Technology center [D]

BFE : Bundesagentur für Arbeit (German work agency) [D]

MBG: Mittelständische Beteiligungsgesellschaft (Public VC) [D]

RWTH Aachen : Rheinisch-Westfälische Technische Hochschule Aachen (School/Institute) [D]

GS: Gründerstipendium : German subsidy for founding a firm [D]

Online Educa : E-learning conference [D]

Regional Cluster
Network : 4
Baden-Württemberg A

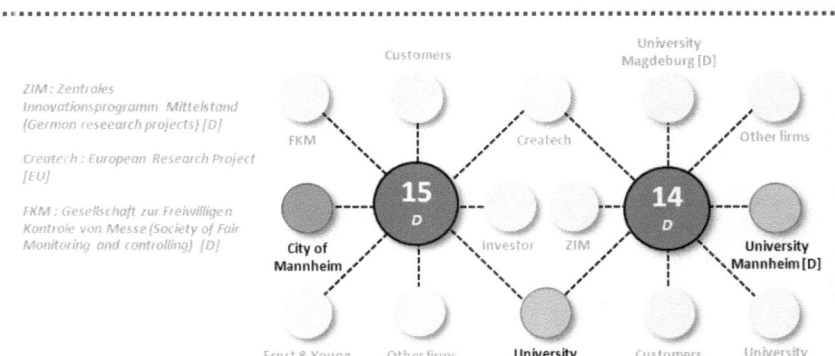

ZIM : Zentrales Innovationsprogramm Mittelstand (German research projects) [D]

Createch : European Research Project [EU]

FKM : Gesellschaft zur Freiwilligen Kontrole von Messe (Society of Fair Monitoring and controlling) [D]

Regional Cluster
Network : 5
Baden-Württemberg B

EEN : *Enterprise Europe Network (information and consultancy network) [EU]*

OSEO : *Financement de l'innovation et de la croissance des PME (Public organization providing strategic and financial support) [F]*

Incuballiance : *Public Innovation Incubator [F]*

Technopole de l'Aube: *Innovation technopole [F]*

CFI : *Centre Francillien de l'innovation (Public innovation promotion agency) [F]*

ADEME : *Agence de l'Environnement et de la Maîtrise de l'Energie (French Energy and Environment public agency) [F]*

AIMA : *Aide à la maturation de projects innovants (NGO promoting innovation) [F]*

ANR : *Agence nationale de la recherche (Public funding for innovation) [F]*

IEF : *Institut d'Electronique Fondamentale (Private research lab) [F]*

CNRS : *Centre national de la recherche scientifique (French national research center) [F]*

Project ACQUEA : *EU partner project [EU]*

CRIT : *Innovation Incubator [F]*

Grand Prix du Concours Lépine : *Innovation contest/fair [F]*

IRD : *Institut de recherche pour le développement (Public innovation promotion agency for Southern regions) [F]*

EDF : *Électricité de France (French public electrical company) [F]*

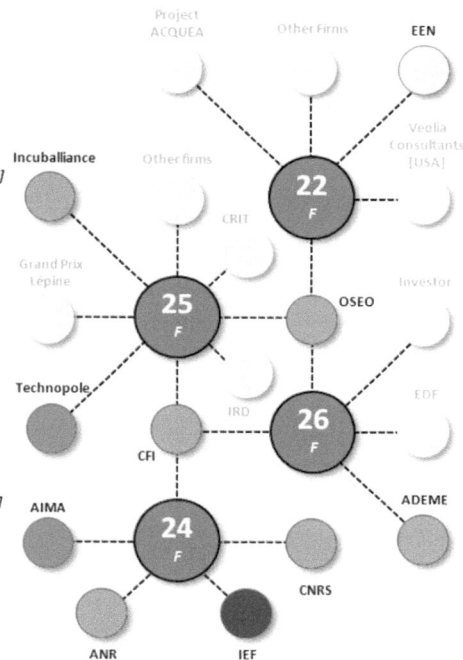

Regional Cluster
Network : 6
Central France / Paris

A3 6 Innovation Networks: International Scope

Figure 15: International Scope Networks. **Source :** Own Illustration

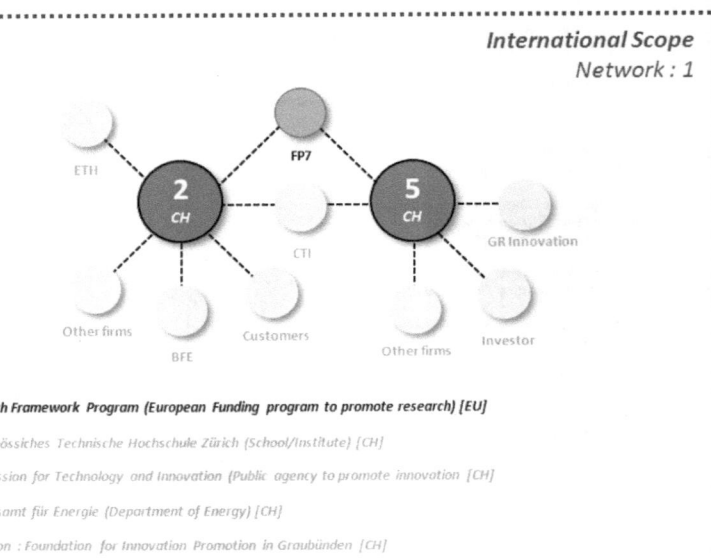

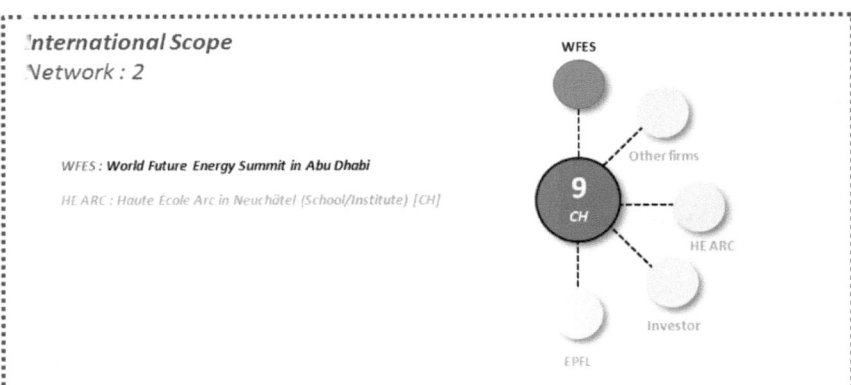

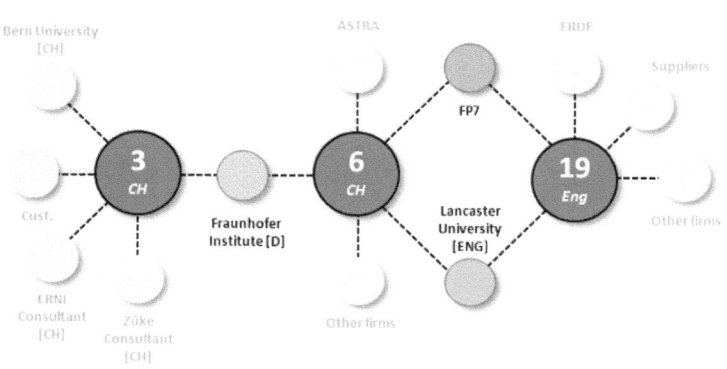

FP7 : Seventh Framework Program (European Funding program to
promote research) [EU]

ASTRA : Bundesamt für Strassen (Federal Roads Office) [CH]

ERDF : European Regional Development Fund [EU]

NHS : National Health Service [UK]

International Scope
Network : 3

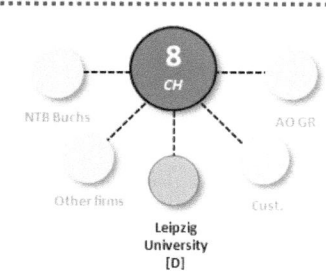

AO GR : AO Foundation Graubünden Research Center [CH]

NTB Buchs : Interstaatliche Hochschule für Technik Buchs
(School/Institute) [CH]

International Scope
Network : 4

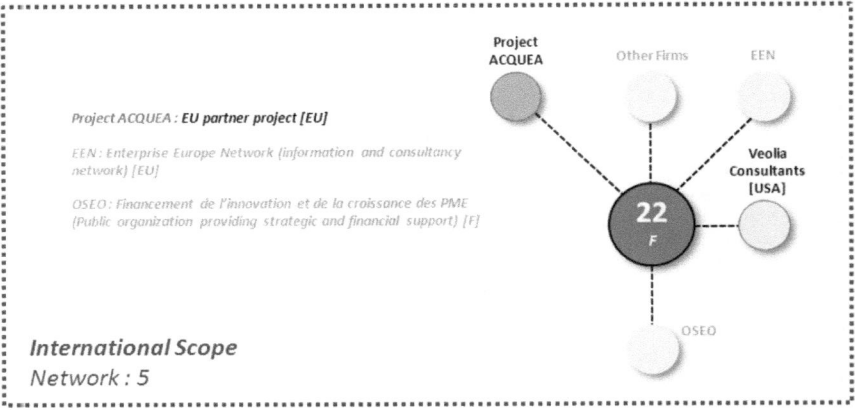

A3 7 Innovation Networks: Isolated Islands

Figure 16: Isolated Islands Networks. **Source :** Own Illustration

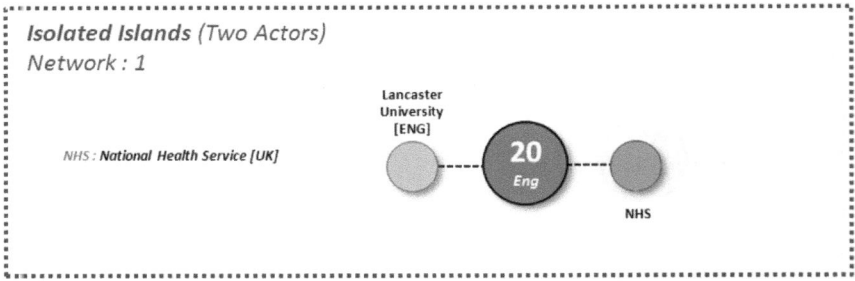

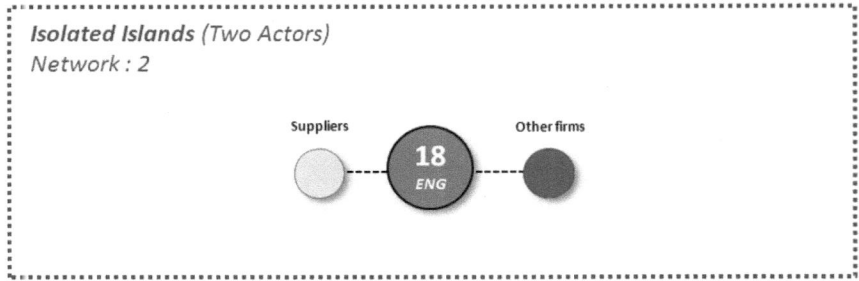

Isolated Islands *(Two Actors)*
Network : 3

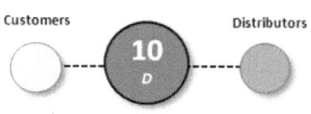

Isolated Islands *(One Actor)*
Network : 4

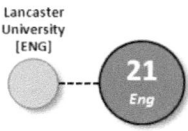

Isolated Islands *(No Actors)*
Network : 5

A4   Data Matrix : Firm Profile

The following charts detail each identified innovation network type and the firms which use them. The firms were categorized based on several individual key network dimensions (see Chapter 5.2). These dimensions were then used to create a general firm profile for each innovation network type.

Table 27: Data Matrix (Firm Profile). **Source :** Own Illustration

Network 1: Knowledge and Learning

Name of Firms is Confidential

Category	1	2	3	4	5	6	7	8	9	10	11	12	13	14
Industry	Motor Technology	Software/Energy	Energy	Energy	Life Science/Medical	Research/Testing	Energy	Life Science/Medical	Transportation	Workshops	IT/Software	IT/Software	IT/Software	Life Science/Medical
Stage	Development	Development	Maturity	Development	Start-up	Maturity	Development	Maturity	Development	Development	Development	Development	Maturity	Development
Size	6-10	11-20	51+	21-50		51+	1-5	21-50	6-10	6-10	6-10	6-10	21-50	6-10
Core Market	Global	Global	Europe-wide	Domestic	Europe-Wide	Global	Domestic	Global	Global				Domestic	Global
Business Model	Product-based	Software	Product-based	Service	Product-based	Service-based	Product-based	Product-based	Product-based	Software	Software	Software	Software	Product-based
Value Chain Process	Production R&D	Production, R&D	R&D	Marketing, R&D	Marketing, distribution, R&D	Production, R&D	Distribution, marketing, production, R&D	Marketing, production, R&D	R&D	Distribution, marketing, production, R&D	Distribution, marketing, production, R&D	Marketing, R&D	Distribution, R&D	Distribution, Marketing, R&D
Motivation to Participate	comp gap, market share	comp gap, market share, finance	comp gap	market share, finance	comp gap, finances	comp gap, market share, finance	comp gap, marketing	comp gap, market share	comp gap, finance	comp gap, market share, finance	comp gap, market share, finance	comp gap, market share, finance	comp gap, finance	comp gap, finance
Timescale	Constant	Constant	Competency need	Project-based	Competency need	Constant	Project-based	Competency need	Project-based	Competency need	Competency need	Competency need	Constant	Project-based
Innovation Formality	informal	formal	informal	informal	formal	formal	informal	informal	informal	informal	informal	informal	formal	informal

Network 1: Knowledge and Learning

Name of Firms is Confidential

Category	15	16	19	20	21	22	23	24	25	26	27	28
Industry	IT/Software	Life Science/Medical	Consulting/Chemical	Consulting/Energy	Horticulture Structures	Energy	Vision Systems, Printing	Coatings/Nano-structures	Transportation	Consulting/Energy	Glass Solutions	Research/Testing
Stage	Development	Start-up		Maturity	Maturity	Maturity	Maturity	Start-up	Development	Start-up	Maturity	Maturity
Size	21-50	11-20	11-20	6-10	21-50	6-10	11-20	1-5	1-5	1-5	21-50	21-50
Core Market	Europe-Wide	Neighbor-countries	Global	Domestic	Domestic	Europe-Wide	Europe-Wide	Global	Europe-Wide	Europe-Wide	Global	Europe-Wide
Business Model	Software	Product-based	Consulting	Consulting	Product-based	Service	Product-based	Product-based	Product-based	Consulting	Product-based	Service
Value Chain Process	Marketing, R&D	R&D	R&D	R&D	R&D	Distribution, R&D	Production, R&D	Distribution, marketing, production, R&D	Distribution, marketing, production, R&D	R&D	Distribution, marketing, R&D	Marketing, R&D
Motivation to Participate	comp gap, market share, finance	comp gap, finance	law, finance	comp gap	comp gap	comp gap, market share, finance	comp gap, market share, finance	comp gap, market share, finance	comp gap, market share, finance	comp gap, finance	comp gap, market share, finance	comp gap, market share, finance
Timescale	Constant	Competency need	Project-based	Project-based	Project-based	Competency need	Project-based	Constant	Constant	Competency need	Competency need	Project-based
Innovation Formality	formal	informal	formal	informal	informal	informal	formal	formal	formal	informal	formal	informal

Network 2: Financial Procurement

Name of Firms is Confidential

Category	2	4	5	6	9	12	13	14	15	19	22	23
Industry	Software/Energy	Energy	Life Science/Medical	Research/Testing	Transportation	IT/Software	IT/Software	Life Science/Medical	IT/Software	Consulting/Chemical	Energy	Vision Systems, Printing
Stage	Development	Development	Start-up	Maturity	Development	Development	Maturity	Development	Development		Maturity	Maturity
Size	11-20	21-50		51+	6-10	6-10	21-50	6-10	21-50	11-20	6-10	11-20
Core Market	Global	Domestic	Europe-Wide	Global	Global			Global	Europe-Wide	Global	Europe-Wide	Europe-Wide
Business Model	Software	Service	Product-based	Service-based	Product-based	Software	Software	Product-based	Software	Consulting	Service	Product-based
Value Chain Process	Production, R&D	Marketing, R&D	Marketing, distribution, R&D	Production, R&D	R&D	Marketing, R&D	Distribution, R&D	Distribution, Marketing, R&D	Marketing, R&D	R&D	Distribution, R&D	Production, R&D
Motivation to Participate	comp gap, market share, finance	market share, finance	comp gap, finances	comp gap, market share, finance	comp gap, finance	comp gap, market share, finance	comp gap, finance	comp gap, finance	comp gap, market share, finance	law, finance	comp gap, market share, finance	comp gap, market share, finance
Timescale	Constant	Project-based	Competency need	Constant	Project-based	Competency need	Constant	Project-based	Constant	Project-based	Competency need	Project-based
Innovation Formality	formal	informal	formal	formal	informal	informal	formal	informal	formal	formal	informal	formal

Network 2: Financial Procurement

Name of Firms is Confidential

Category	24	25	26	27	28
Industry	Coatings/Nano-structures	Transportation	Consulting/Energy	Glass Solutions	Research/Testing
Stage	Start-up	Development	Start-up	Maturity	Maturity
Size	1-5	1-5	1-5	21-50	21-50
Core Market	Global	Europe-Wide	Europe-Wide	Global	Europe-Wide
Business Model	Product-based	Product-based	Consulting	Product-based	Service
Value Chain Process	Distribution, marketing, production, R&D	Distribution, marketing, production, R&D	R&D	Distribution, marketing, R&D	Marketing, R&D
Motivation to Participate	comp gap, market share, finance	comp gap, market share, finance	comp gap, finance	comp gap, market share, finance	comp gap, market share, finance
Timescale	Constant	Constant	Competency need	Competency need	Project-based
Innovation Formality	formal	formal	informal	formal	informal

Network 3: Public - Private Cooperation

Name of Firms is Confidential

Category	1	2	3	4	5	6	7	8	11	12	13	14
Industry	Motor Technology	Software/Energy	Energy	Energy	Life Science/Medical	Research/Testing	Energy	Life Science/Medical	Webshop	IT/Software	IT/Software	Life Science/Medical
Stage	Development	Development	Maturity	Development	Start-up	Maturity	Development	Maturity	Development	Development	Maturity	Development
Size	6-10	11-20	51+	21-50		51+	1-5	21-50	6-10	6-10	21-50	6-10
Core Market	Global	Global	Europe-wide	Domestic	Europe-Wide	Global	Domestic	Global			Domestic	Global
Business Model	Product-based	Software	Product-based	Service	Product-based	Service-based	Product-based	Product-based	Software	Software	Software	Product-based
Value Chain Process	Production, R&D	Production, R&D	R&D	Marketing, R&D	Marketing, distribution, R&D	Production, R&D	Distribution, marketing, production, R&D	Marketing, production, R&D	Distribution, marketing, production, R&D	Marketing, R&D	Distribution, R&D	Distribution, Marketing, R&D
Motivation to Participate	comp gap, market share	comp gap, market share, finance	comp gap	market share, finance	comp gap, finances	comp gap, market share, finance	comp gap, marketing	comp gap, market share	comp gap, market share, finance	comp gap, market share, finance	comp gap, finance	comp gap, finance
Timescale	Constant	Constant	Competency need	Project-based	Competency need	Constant	Project-based	Competency need	Competency need	Competency need	Constant	Project-based
Innovation Formality	informal	formal	informal	informal	formal	formal	informal	informal	informal	informal	formal	informal

Network 3: Public - Private Cooperation

Name of Firms is Confidential

Category	15	16	19	20	21	22	23	24	25	26	27	28
Industry	IT/Software	Life Science/Medical	Consulting/Chemical	Consulting/Energy	Horticultural Structures	Energy	Vision Systems, Printing	Coatings/Nano-structures	Transportation	Consulting/Energy	Glass Solutions	Research/Testing
Stage	Development	Start-up		Maturity	Maturity	Maturity	Maturity	Start-up	Development	Start-up	Maturity	Maturity
Size	21-50	11-20	11-20	6-10	21-50	6-10	11-20	1-5	1-5	1-5	21-50	21-50
Core Market	Europe-Wide	Neighbor-countries		Domestic	Domestic	Europe-Wide	Europe-Wide	Global	Europe-Wide	Europe-Wide	Global	Europe-Wide
Business Model	Software	Product-based	Consulting	Consulting	Product-based	Service	Product-based	Product-based	Product-based	Consulting	Product-based	Service
Value Chain Process	Marketing, R&D	R&D	R&D	R&D	R&D	Distribution, R&D	Production, R&D	Distribution, marketing, production, R&D	Distribution, marketing, production, R&D	R&D	Distribution, marketing, R&D	Marketing, R&D
Motivation to Participate	comp gap, market share, finance	comp gap, finance	law, finance	comp gap	comp gap	comp gap, market share, finance	comp gap, market share, finance	comp gap, market share, finance	comp gap, market share, finance	comp gap, finance	comp gap, market share, finance	comp gap, market share, finance
Timescale	Constant	Competency need	Project-based	Project-based	Project-based	Competency need	Project-based	Constant	Constant	Competency need	Competency need	Project-based
Innovation Formality	formal	informal	formal	informal	informal	informal	formal	formal	formal	informal	formal	informal

Network 4: Vertical Integration

Name of Firms is Confidential

Category	1	2	3	4	7	8	10	13	14	15	16	18
Industry	Motor Technology	Software/Energy	Energy	Energy	Energy	Life Science/Medical	Workshops	IT/Software	Life Science/Medical	IT/Software	Life Science/Medical	Energy
Stage	Development	Development	Maturity	Development	Development	Maturity	Development	Maturity	Development	Development	Start-up	Development
Size	6-10	11-20	51+	21-50	1-5	21-50	6-10	21-50	6-10	21-50	11-20	
Core Market	Global	Global	Europe-wide	Domestic	Domestic	Global		Domestic		Europe-Wide	Neighbor countries	Domestic
Business Model	Product-based	Software	Product-based	Service	Product-based	Product-based	Software	Software	Product-based	Software	Product-based	Consulting
Value Chain Process	Production, R&D	Production, R&D	R&D	Marketing, R&D	Distribution, marketing, production, R&D	Marketing, production, R&D	Distribution, marketing, production, R&D	Distribution, R&D	Distribution, Marketing, R&D	Marketing, R&D	R&D	distribution, R&D
Motivation to Participate	comp gap, market share	comp gap, market share, finance	comp gap	market share, finance	comp gap, marketing	comp gap, market share	comp gap, market share, finance	comp gap, finance	comp gap, finance	comp gap, market share, finance	comp gap, finance	comp gap
Timescale	Constant	Constant	Competency need	Project-based	Project-based	Competency need	Competency need	Constant	Project-based	Constant	Competency need	Project-based
Innovation Formality	informal	formal	informal	informal	informal	informal	informal	formal	informal	formal	informal	informal

Network 4: Vertical Integration

Category

Category	19	20	23	27	28
Industry	Consulting/Chemical	Consulting/Energy	Vision Systems; Printing	Glass Solutions	Research/Testing
Stage		Maturity	Maturity	Maturity	Maturity
Size	11-20	6-10	11-20	21-50	21-50
Core Market	Global	Domestic	Europe-Wide	Global	Europe-Wide
Business Model	Consulting	Consulting	Product-based	Product-based	Service
Value Chain Process	R&D	R&D	Production, R&D	Distribution, marketing, R&D	Marketing, R&D
Motivation to Participate	law, finance	comp gap	comp gap, market share, finance	comp gap, market share, finance	comp gap, market share, finance
Timescale	Project-based	Project-based	Project-based	Competency need	Project-based
Innovation Formality	formal	informal	formal	formal	informal

Network 5: Regional Clusters

Name of Firms is Confidential

Category	1	2	5	7	11	12	13	14	15	16	19	20
Industry	Motor Technology	Software/Energy	Life Science/Medical	Energy	Webshops	IT/Software	IT/Software	Life Science/Medical	IT/Software	Life Science/Medical	Consulting/Chemical	Consulting/Energy
Stage	Development	Development	Start-up	Development	Development	Development	Maturity	Development	Development	Start-up		Maturity
Size	6-10	11-20		1-5	6-10	6-10	21-50	6-10	21-50	11-20	11-20	6-10
Core Market	Global	Global	Europe-Wide	Domestic			Domestic		Europe-Wide	Neighbor-countries	Global	Domestic
Business Model	Product-based	Software	Product-based	Product-based	Software	Software	Software	Product-based	Software	Product-based	Consulting	Consulting
Value Chain Process	Production, R&D	Production, R&D	Marketing, distribution, R&D	Distribution, marketing, production, R&D	Distribution, marketing, production, R&D	Marketing, R&D	Distribution, R&D	Distribution, Marketing, R&D	Marketing, R&D	R&D	R&D	R&D
Motivation to Participate	comp gap, market share	comp gap, market share, finance	comp gap, finances	comp gap, marketing	comp gap, market share, finance	comp gap, market share, finance	comp gap finance	comp gap, finance	comp gap, market share, finance	comp gap, finance	law, finance	comp gap
Timescale	Constant	Constant	Competency need	Project-based	Competency need	Competency need	Constant	Project-based	Constant	Competency need	Project-based	Project-based
Innovation Formality	informal	formal	formal	informal	informal	informal	formal	informal	formal	informal	formal	informal

Network 5: Regional Clusters

Name of Firms is Confidential

Category	21	22	24	25	26	27	28
Industry	Horticultural Structures	Energy	Coatings/Nano-structures	Transportation	Consulting/Energy	Glass Solutions	Research/Testing
Stage	Maturity	Maturity	Start-up	Development	Start-up	Maturity	Maturity
Size	21-50	6-10	1-5	1-5	1-5	21-50	21-50
Core Market	Domestic	Europe-Wide	Global	Europe-Wide	Europe-Wide	Global	Europe-Wide
Business Model	Product-based	Service	Product-based	Product-based	Consulting	Product-based	Service
Value Chain Process	R&D	Distribution, R&D	Distribution, marketing, production, R&D	Distribution, marketing, production, R&D	R&D	Distribution, marketing, R&D	Marketing, R&D
Motivation to Participate	comp gap	comp gap, market share, finance	comp gap, market share, finance	comp gap, market share, finance	comp gap, finance	comp gap, market share, finance	comp gap, market share, finance
Timescale	Project-based	Competency need	Constant	Constant	Competency need	Competency need	Project-based
Innovation Formality	informal	informal	formal	formal	informal	formal	informal

Network 6: International Scope

Name of Firms is Confidential

Category	2	3	5	6	8	9	19	22	27	28
Industry	Software/Energy	Energy	Life Science/Medical	Research/Testing	Life Science/Medical	Transportation	Consulting/Chemical	Energy	Glass Solutions	Research/Testing
Stage	Development	Maturity	Start-up	Maturity	Maturity	Development		Maturity	Maturity	Maturity
Size	11-20	51+		51+	21-50	6-10	11-20	6-10	21-50	21-50
Core Market	Global	Europe-wide	Europe-Wide	Global	Global	Global	Global	Europe-Wide	Global	Europe-Wide
Business Model	Software	Product-based	Product-based	Service-based	Product-based	Product-based	Consulting	Service	Product-based	Service
Value Chain Process	Production, R&D	R&D	Marketing, distribution, R&D	Production, R&D	Marketing, production, R&D	R&D	R&D	Distribution, R&D	Distribution, marketing, R&D	Marketing, R&D
Motivation to Participate	comp gap, market share, finance	comp gap	comp gap, finance	comp gap, market share, finance	comp gap, market share	comp gap, finance	law, finance	comp gap, market share, finance	comp gap, market share, finance	comp gap, market share, finance
Timescale	Constant	Competency need	Competency need	Constant	Competency need	Project-based	Project-based	Competency need	Competency need	Project-based
Innovation Formality	formal	informal	formal	formal	informal	informal	formal	informal	formal	informal

Network 7: Isolated Islands

Name of Firms is Confidential

Category	10	17	18	20	21
Industry	IT/Software	Lab	Consulting	Consulting/Energy	Horticultural Structures
Stage	Maturity	Development		Maturity	Maturity
Size	51+	1-5	1-5	6-10	21-50
Core Market	Europe-Wide	Domestic	Domestic	Domestic	Domestic
Business Model	Software	Service	Consulting	Consulting	Product-based
Value Chain Process	Production, R&D		R&D	R&D	R&D
Motivation to Participate	comp gap		comp gap, finance	comp gap	comp gap
Timescale	Project-based		Project-based	Project-based	Project-based
Innovation Formality	informal	informal	informal	informal	informal

A5   Interview Guide

The following is the interview guide used by the KARIM researchers in conducting their interviews. The guide was prepared to establish a thread of consistency binding the different interviewers, firms, and countries. The interview guide was distributed in the Action 1: Interactive Innovation MAP Handbook for Data Collection (Becker & Bau, 2011).

Interview Guidelines

Firm/Institution:

Name, First Name: ....

Function/Job: ....

Nationality: ....

E-Mail: ....

Telephone: ....

Date of Interview: ....

Place of Interview: ....

Time of Interview-start: ....

Time of Interview-finish: ....

Interviewer: ....

Interview Guidelines

"North West Europe Knowledge Transfer Network
– "Development of an Interactive Innovation Map" -

→ *Greeting and Thanks for Participation*
→ *Asking permission to record the interview*
→ *Clarification of the Interview Proceedings*

Preliminary Notes

- Initial Greeting and expressing gratitude for participation
- Can the interview discussion be recorded? All data will remain with us.
- Presentation of SIFE, HTW Chur, Swiss Partner with respect to the EU-Project KARIM (Knowledge Acceleration Responsible Innovation Meta-network)
- 8 partners from 7 countries within Nord-West-Europe; (CH, F, D, NL, Wales, England)
- The aim of the Project is to significantly increase the access of SMEs throughout North West Europe to valuable innovation and technology thus enhancing and sustaining their world-wide competitive potential.
- Project Objectives:
  - Increased cutting-edge technology transfer and access to innovation support for KMUs (regional Swiss SME association)
  - Facilitation and enhancement of transnational co-operation and net-working
  - Supporting and assisting KMUs with transnational Innovation projects
- Interview Objectives:
  - How does Innovation develop from the perspective of the firm, including key-activities, influencing factors, etc.?
  - Determination of specific interests and experience to-date
  - Development of a real-time innovation activity map based on actual innovation projects on a case by case analysis; (not a telephone book!)
  - Innovation-map Target Group: Firms that seek co-operative partners; innovation support organisations.
- Potential advantage for Interview partner/firm/approach:
  - Profiling and Image cultivation e.g. creation of a template of Best Practice Examples.
  - Prior to publishing the interviewee will have the opportunity to both review the content and voice their opinion as well as being able to Veto any part of the content and or remain an anonymous participant.

Interviewstrategy:

- Create an open atmosphere
- Use the questions in the guideline to structure the interview but do not necessarily follow the structure. However, check the guideline at the end if all questions are answered implicitly or explicitly.
- Try to get specific examples, ideally of transnational cooperative innovation projects and try to refer to those examples throughtout the interview.
- Identify the key actors in innovation support. We do not want to know with whom they chaned business cards but who really boosted their innovation projects.

Part 1: Questions for the Firm and the interview Partner

Initial Phase: Confirmation and up-dating of the previously researched information relating to the firm and interview partner.

As much of the basic detailed information can be easily accessed in the internet, it may be useful to concentrate on the main focus areas of the interview in order to achieve a clear and common base on which to conduct the interview.

(1) What is your core product or business?

(2) What is the target market or country for your business?

If your firm is already dealing internationally, which market or country has significant relevance or importance for your business?

(3) Could you describe briefly your duties and responsibilities within the firm? In addition, how are the tasks and responsibilities distributed within the management team?

(4) What is your role or involvement in the innovations process of your firm?
For example, facilitating co-operation, development, decision making?

(5) If no researchable data is available, note firm size, employee number, and year of founding

Part 2: Questions regarding the Innovation Process

(6) We would like to begin with a number of questions relating to innovation related activity within your firm.

(7) What relevance has Innovation within your firm?

(8) How effective are the innovation related activities within your firm?
    a. Have you established a formal innovation process procedure? Has this process changed since the founding of the company, e.g. in consequence of growth phases, etc.?
    b. Do you have a management system and working protocol for handling and benefitting from new ideas?
    c. Do you have a clear and traceable allocation and distribution of resources, including personnel, finance, etc. to innovation projects?
    d. Is there an effective priority system in place to handle different projects managed at the same time? Is it clear and transparent in the case of parallel running projects which should take precedence?
    e. How many innovation projects are you running currently?
    f. Is there a clear and transparent assignment of responsibility for innovation project management?
    g. Does your organisation's culture actively support initiatives for innovative projects?
    h. Is the top management layer of your company actively interested and supportive of innovation projects?

(9) Do you have a specific example of an actual Innovation project, ideally one with international links?
[If possible, try to select one specific project example for the following questions.

(10) Could you give us a brief description of this project?(Process, Theme, Product, Technology, Partner, Market, Project budget, percentage of sales/turnover)

Part 3: Questions concerning the internal & external impetus for innovation

With respect to the specific project example discussed, we would now like to ascertain the key actors behind the process and specifically to address the main positively influencing factors in its development.

(11) How was the project idea generated and how was it developed?

(12) Which steps were taken to ensure the operational realisation of the project after the project idea, aim and objectives were defined?

(13) Who were the most important internal and external stakeholders in the project? (**Internal**: R & D Manager, Owners, Managing directors, Employees; **external**: Investors, Educational Institutions, Development Partners, Customers, Local communities, General public, Suppliers and Logistics)? (Not those exchanging a business card but influential)

(14) How did the contact to these project stakeholders arise?

(15) How were these stakeholders integrated and relationship defined? (which form or type of partnership?)

(16) What benefit have these participants brought to the project or to your firm?

(17) If you perceive that no benefits were observable then why? What was missing? And what in your opinion is required?

(18) In your opinion, which internal factors have had a positive influence on the innovations process? Could you outline this aspect for me in more detail on the basis of the example that you have previously described? (Company culture, employee skills and qualifications, organisation, resources,...)

(19) Which external factors have a positive effect on the internal innovation process within the company? Do you also have any examples; ideally referring to the same project. (Research funding, Political decisions, Economic situation, technological developments, Competitive environment, ...)

(20) How did you finance this type of innovation project?

(21) From your point of view, have you experienced any unforeseen problems with one of you innovation projects after the innovation has been introduced on the market? If yes, which ones? For which reasons?

(22)I think there is one question missing in terms of unforeseen problems with the innovation once its on the market. This question can pitch both how good the company managed risks during the development process and how wide the company considers its impacts.

Part 4: Questions relating to Co-operation and Partnerships

In the next block of questions we will deal with the central theme of the Project, namely discussion of co-operation and partnerships.

(23) What were your expectations in relation to co-operation in the innovation process?

(24) What support or co-operation have you experienced in relation to actual innovation projects? (For example, educational institutes, clients, customers, firms, etc.)

(25) Has the co-operation mentioned above been instrumental in the successful attainment of the aim and objectives of your innovation project?

(26) With respect to specific innovation projects, do they normally involve a significant contribution of international co-operation?

(27) In general or related to the before mentioned project, what for you are the crucial factors relating to the building and implementation of co-operation and joint ventures and do you have any specific examples?

(28) Which Partner (category) has in your opinion had the most significance?

Part 5: Summary Conclusion of Interview

We would like to summarise and highlight the key points discussed within the interview; with specific reference to international co-operation, as well as 'Responsible Open Innovation'. (Keep in mind the story line)

(29) From your point of view, what are the key aspects and critical factors influencing the composition and implementation of innovation projects in general and do you have any related examples?

(30) From your perspective, could you summarize who the essential key actors are with respect to all the projects that you are involved with?

(31) 'Responsible Open Innovation' is the essence and focus of our EU- Project
   a. What would you consider to be Responsible Innovation?
   b. What does the term 'Open Innovation' mean for you?
   c. What relevance have the aforementioned factors for you?

(32) Does the project region (NEW) and related theme hold any interest or significance for you?

(33) The common aim of this EU-Project is the establishment of transnational co-operation within the NWE region. Therefore, it is important for us to ascertain what support you expect from the political system when deciding to enter a market or form co-operative working relationships with enterprises within one of the 'NWE' countries

(34) Have you any further comments or points to add?

- *Thank-you very much for taking the time to share with us your personal experience and knowledge.*
- → *We would appreciate it, if we could receive feedback with respect to the first draft of this interview and information handling. Furthermore, we would like to remain in contact and inform you of the developments in the innovation map with any additional input. The time frame will be at the end of this year at the earliest, but more probably during the second half of 2012.*
- → *Conclusion of Interview*